MW01518317

To: Pattie

Thank you for your
support!

Happy Reading

[signature]

When I Died

Elizabeth Eckert

WHEN I DIED

Copyright © 2016 Elizabeth Eckert.

*All rights reserved. No part of this book may be used or reproduced by any means,
graphic, electronic, or mechanical, including photocopying, recording, taping or by
any information storage retrieval system without the written permission of the author
except in the case of brief quotations embodied in critical articles and reviews.*

*This is a work of fiction. All of the characters, names, incidents, organizations, and dialogue
in this novel are either the products of the author's imagination or are used fictitiously.*

iUniverse books may be ordered through booksellers or by contacting:

iUniverse
1663 Liberty Drive
Bloomington, IN 47403
www.iuniverse.com
1-800-Authors (1-800-288-4677)

*Because of the dynamic nature of the Internet, any web addresses or links contained in
this book may have changed since publication and may no longer be valid. The views
expressed in this work are solely those of the author and do not necessarily reflect the
views of the publisher, and the publisher hereby disclaims any responsibility for them.*

*Any people depicted in stock imagery provided by Thinkstock are models,
and such images are being used for illustrative purposes only.
Certain stock imagery © Thinkstock.*

ISBN: 978-1-4917-7912-5 (sc)
ISBN: 978-1-4917-7911-8 (hc)
ISBN: 978-1-4917-7910-1 (e)

Library of Congress Control Number: 2016915539

Print information available on the last page.

iUniverse rev. date: 07/25/2016

This book is dedicated to my precious children;

Dierdre, for critiquing my first published book and for being the most beautiful, independent young woman I know.

Michael, for being determined enough to be who he is… an outstanding and courageous man.

J.J., for always adding humour, support, and clarity to my daily life.

What an honorable and amazing man you are becoming.…

Anakin, for showing inconceivable inspiration, heart, and strength; despite everything, you have endured. My forever little man.

And to Tara-Jade Scarlett, you're not with us in body but your spirit is our angel, and on your wings our hearts soar.

To my Mom, Winnie and my Father, Colin (In Heaven);

Thank you for believing in me and encouraging my dream into a reality.

*Edited by Dierdre Eckert and Jonathan Entwisle.

1

My death and the gift

Death. We all think about it, dream about it, fantasize about it, and even romanticize it a little. I most certainly did. It's all in good fun. At seventeen, you don't seriously contemplate dying - like really, truly being stone-cold dead - although there have been times in my short life where I let my imagination run wild about stupid ways to die to pass the time, like having a rare form of cancer or being eaten by tigers. I mainly wanted to distract myself from other more imminent problems and fears (teens are so morbid). I can't remember though; did I imagine that there would be much pain, the bright light at the end of the tunnel, or seeing dead loved ones? Whatever I thought before was not even close to this. The fact that I am actually thinking these thoughts at the same time I am sure I am dying is a mystery to me. No logic found, no explanations.

There is no tunnel yet, if there is to be one, but there are lights. I can see flashing red and blue globes swirling around in a monotonous dance. I can feel the winter breath on my pale skin and the damp cold of the pavement underneath my body. There are people talking. Both male and female voices are uttering indefinable phrases. Perhaps I don't want to understand what they are saying. I wonder if I am supposed to experience pain now; I did a few minutes ago. Or was that hours ago? I'm not sure. It seems like just a few minutes ago, that I was staring up at the sky, the white, fluffy flakes of snow falling softly on my face, waiting patiently for the bus. I'm confused that I can still sense my body even though I am staring down at myself from an unknown height. I think the sensation is fading though; I am not as cold anymore. Is this astral projection? I read about that once. I'm positive I am dying, or already dead. I continue to watch, detached, fear and panic rising as suited men come to wrap my lonely body in a black bag. I've seen this happening in the movies, but now it is happening to me. Faintly, I sense their hands on my cold skin. My long, unruly black hair catches in the zipper, though there is no tug on my scalp. Inconceivably, it is happening, I am leaving my body. Resignation replaced my fear slowly as I look away from the grotesque body bag that holds the figure I have lived in for a mere seventeen years to see the faces of people gathered at my death scene.

The police officers are in work mode. They are not unsympathetic but solemnly submissive to the task. Parents, my parents, must be notified. The paramedics have completed their duty. No resuscitation needed. I am long past dead, no hope for return. I briefly note a few onlookers. With surprise, I realized it did not bother me that they are curious. I would have thought that I would be angry or embarrassed that strangers were staring at my body, my corpse lying pale and vacant on a sidewalk (and I look like crap), just a few feet away from the city transit bus stop on this gray November Saturday afternoon. Do they wonder how I died? I wonder too.

My day started without any premonition of my untimely demise, without even a hint that the useless last words I said to my mother would be the final words I would speak while living. I awoke late, as I often did on a weekend. Reaching over, I slammed my snooze button for the third time, wishing I might sleep all day. My pink-painted room, decorated with light pine furniture and an undetermined number of stuffed animals, was like a jail cell these last few months. I tolerated the enclosure only to sleep and because it offered a facade of solitude and avoidance of the quiet chaos outside my walls. Each day, my formerly friendly sanctuary became more like a steel cage that bound and trapped me away from my 'real' happiness. The faces on my wall, of the Jonas brothers, Edward and Bella and Gerard Butler, began to stare back at me with looks of disapproval and contempt.

They seemed to be judging me for my pessimistic and desolate thoughts. (I think I was going crazy).

I jumped out of bed, dressing for the predictable cold November day in the prairies. With talented speed, I plastered makeup around my light-brown eyes, high cheekbones, and thinner-than-I-would-like lips. I left my hair as it always was, doing its thing, hanging in stubborn turmoil around my shoulders. I paused for a moment in front of my full-length mirror and wondered if I could be putting on a bit of weight. My size seven blue jeans seemed a little tighter than before (I am still kind of hot though). Thankfully, the mauve long-sleeved scoop-neck top my step-dad had bought me on his last trip out of town still flattered my slim figure. It had fast become my favourite with each passing compliment I received while wearing it.

How silly to have been so critical of my appearance when a few short hours later the color would be drained from my face and my once-sparkling brown eyes would become empty, the shade of black coal in the light of the wintery sky. My grandma always said if she had one super power it would be the power of foresight. I now understand what she meant. If I had known I'd be dead this afternoon, I might have smiled when I left the house today. I could have sneaked a peek to see my little sister playing with her Barbie's or coloring a picture like seven-year-olds are supposed to do. I would have kissed her beautiful brown hair, tousled her soft curls, tied the cloth belt on the pretty dress she was bound to be wearing. I would have told her that, in a few more years, she would be free to leave this place and put her lonely, disturbing childhood behind her. But I didn't. I didn't even open her door, or stop to listen for her high, gentle laugh. In ignorance, I stomped past her little purple room, past the pale-yellow bathroom we shared. Not pausing at my mom and dads artistically decorated, eclectic room, I trudged down the beige plush-carpeted stairs to the modernized kitchen, where I would be just a jump away from my escape. I yelled goodbye to my mom, not caring if she heard me. We shared an understanding: I came and went as I pleased, called if I would be out past ten, only drank alcohol if I was at home (as far as she knew), and took care of my sister 3 nights a week to pay for my cell phone bill. Simple, unrestrictive, detached.

Lost in thought (as improbable as that seems), I did not notice when they took my body. I didn't even feel it go. The scene below changed so drastically that I almost wondered if I could be just dreaming and this was one of those twists or scene changes that so often occur in your nightmares, like after eating pizza or rocky road ice cream late at night. No, I wasn't dreaming. I was dead and unconnected altogether now from my body. I started to panic again. Where was I? I can see everything around me, although I appear to have no body, no form. What am I now? Am I a spirit, a ghost, or the beginnings of a disgusting, rotting zombie? How can I be seeing, thinking, or knowing anything? I am dead, dead, dead.

The paramedics were gone, although the investigators were still taking pictures, notes, hair, and dirt samples. The police sectioned off my crime scene. But there was no crime. I did not get mugged, struck down, shot, or stabbed. I just died. People were still staring at the last place I had stood with expressions of sadness, surprise, question, and curiosity. One woman stared more intently than the others, looking almost angry at the scene before her. She looked up to the sky, said what appeared to be a prayer, and turned towards my spiritual location. She then did something unexpected: she smiled, nodded her head, and walked away. It was as if she sensed I was there. Could I follow her, communicate with her to ask even one of the million questions I had racing in my mind?

I attempted to force myself to move, my mind stretched towards legs that were no longer there. I saw the large gray-haired woman turn the corner at the end of the street, but I could not follow. Despair encompassed my soul as I envisioned myself being stuck in this one spot for the rest of my...what? Existence? I couldn't make myself move or even imagine myself somewhere else. I needed to think; that, at least, I was obviously still capable of doing. There must be a way to not be here, somewhere in the atmosphere just above this isolated street in the middle of the city I grew up in. I did not want be bound to this location forever.

After taking in my surroundings, I found it to have changed yet again. The sun left a faint glow, the street was empty, although my death-zone tape still hung, and I was still dead. Snow fell as softly as cotton balls to the wet ground where I had spent my last living moments, where I had left the number 21 bus in hopes of catching the 43 down to the mall, where I was planning to meet up with my best friend, Audrey. We needed to discuss plans for our Saturday night.

Audrey would call my mom to ask where I was after texting my phone a million times (she was highly persistent when she wanted to be). She would be worried about me and think I had ditched her. Mom would tell her to give me time, that I must have missed a bus or two, and I'd be along shortly. Hours would pass before the police would go to my house, on the quiet little nothing-ever-happens-here cul-de-sac where we live. Where I lived no more. Steph, my sweet little sister, would listen behind a door or around a corner, hiding. She would hear a stranger in a uniform tell my parents I was now dead of unknown causes. On the other hand, it was possible they knew something I didn't; perhaps they knew how I died. Perhaps only an autopsy would show, to all but me, I had an undetected brain tumour, an unknown heart condition, or an unheard-of disease that took my life before I had a chance to live. I had aspirations to fulfil, obstacles to overcome, places I wanted to see, and people I had yet to meet who would offer revelations to change my cynical outlook on love and relationships in general.

My mom, Heather Mackenzie (the former Heather Shaw, Heather Lewis, and Heather Taylor), would cry and then grab a drink from her not-so-private stash in the top of the pantry cupboard. After a long, undiluted drink of courage, she would ask, "Was it suicide? Drugs? Was it that 19-year-old that she was dating for a few weeks - what was his name, Steve? Shawn?" Her flushed face would add humour to the heavy makeup she wore to hide the scars of childhood acne and years of unbridled alcohol indulgence. Her immaculately kept, dyed-blonde hair would fall out of the elaborate pinned-up style that she fussed over each day. Her brown eyes would be red rimmed and smudged with black eyeliner and mascara. She would stand tall with each of the 5 Ft. 11 inches that heredity had given her, not slouching even so much as to wrinkle one of the pressed, expensive Armani dress suits she practically lived in.

A complete contradiction would be my stepfather, Jacob Mackenzie, a rather genteel addition to my mother's long list of husbands. He would stare quietly and allow my mother to ramble. Then he would become angry with himself for not "seeing any signs." Jacob was a spiritual man. He would demand to see my body to be sure that I was dead, and that there was no run-away cover-up attempt. Jacob tended to be a tad paranoid. After he calmed down (his temper was quick but over as fast), he would wrap his muscular arms around my mother, ruffling her perfect clothes. He would tuck her head under his chin, despite her height, and deliver words of comfort. Another gray hair would add itself to the many that were appearing at the temples of his buzzed black hair.

Many thought that Jacob was my biological father due to us having the same hair color and large bones, incongruous with the slim physique we both graced. Although Jacob had only been in our lives for 8 years, he was as close to a father as I would ever have. Neither he nor my Mother would spare a moment to consider that Steph might be lurking nearby, hearing this horrible news through the lips of outsiders, not with the tenderness she would need and deserve. How do I know this so certainly? That was the scene three years and two weeks before, when police had come to our same mahogany front door, to tell my parents about Kieran; I was the hidden girl then listening to the report of death in my family.

I wondered how my parents would adapt to yet another shocking loss inflicted on this already troubled, sad household. First my older Brother Kieran, now me; stubborn, friendly, loving, little Adrianna, gone. I was somewhat (downright morbidly) curious to know how everyone would react to my being dead. Intrigued about the feelings of my family, my friends, I let my mind imagine …OH! My funeral! Horrible! Visualizing what it would be like to be there to watch the tragedy hit home made me want to bawl like a baby. Regret tore me apart. In hindsight, I should have tried to help more, pull us all together, instead of always wanting to escape. It should have been my job to ensure that Steph would have someone to help guide her, teach her that the world did not have to be lonely and difficult, and then to be there to give her the push to leave and go on to have a decent life on her own.

* * * * *

"She's not ready, son. You must allow the process to go on as it must; pushing things will only serve to hurt her, and you'll run the risk of her being lost forever. Be patient," the gray-haired man stated sternly.

"Listen, Shadow, I…I mean, we have been waiting patiently for an unbearably long time, haven't we?" he stammered. "Regardless, I know the amulet will bring her closer. Can you grant me permission to at least try?" the young man more or less pleaded.

Making eye contact with the old man was unsettling. His heavily lined face was almost comical in comparison to his youthful, pale, sea-blue eyes and tall, lean, but sturdy stature.

"You have my permission but also my warning. The charm will raise questions and may not serve the purpose you intend. We have only used the amulet with the ones that are 'special' to help them to become cognizant of their powers and to speed up the process of their development. It is primarily used for those who have already acknowledged and accepted their fate. They have all moved forward now. My boy, time will allow us all what we have been waiting for. If she is to be the one, then let the winds of fate blow accordingly."

Shadow turned with a whisper of his black cloak and disappeared as quickly as he had appeared only moments before. The young man gazed unseeing after the wise old man left. He wished he could argue, wished he could prove what he felt to be so true, that she was the one. 'Time, what a strange concept, what a useless measure,' he thought. Bitterness crept back, threatening to overflow, and take control again. Shutting it out quickly, he smiled weakly. 'Yes, the process will go on at the pace it is meant to, with my help. Then I can finally tell her how crucial she is...to everything.'

To stand next to a person who does not know you are there can be a terrifically humorous situation. Amazing how many souls, both living and dead, can invade the same room without any disastrous consequences or even awareness, apart from the occasional goose bumps and hair-on-the-neck raising. Hundreds of complex, infinitesimal energy forms manipulating the same environment, same space, in an organized, methodical fashion. It truly was quite extraordinary. He understood why not all who died could linger to visit the living world once again. With so many more energy forms, manoeuvring would not be possible.

There were many people here in the large entrance to the Foothills hospital. After the smell of extreme cleanliness, the aroma of coffee delightfully flooded his nose. It was a marvellous scent. The little cafe stood just off to his right across from the elevators. Every few seconds pages, codes, and announcements rang for Doctors, patients, families, and porters. It was a busy place. In a city of over one million people, there were tragic events happening all the time: car crashes, heart attacks, the seasonal flu, and a few murders.

Delving into the unsuspecting minds of each stranger was only possible with the amulet in his pocket. It enhanced certain traits one possessed, it seemed. His ability to invade thoughts was usually only centered on one person at a time, if at all. It was more of a heightened perception of thoughts than a mind-reading technique. It was quite intriguing how the amulet allowed his latent abilities to flourish. He rather liked it.

A tall, pretty, brunette standing next to him had the opinion that winter would slam 3 feet of snow in the next few days to increase the swamping in the emergency room. She was right. Apparently, the amulet brought out the weatherman in him as well. He looked around for a moment to recall the visual aspect of his memory of the last time he was here; he followed the same route he had in the past, although under very different circumstances. This was the last place he had been 'physically' anywhere. How it seemed so long ago. Of course, much had changed. Completed and ongoing renovations were apparent here and in hundreds of other places that he had been and not been before. The city had almost quadrupled in size, and technology had flourished. How the world had changed.

He made his way down the vaguely familiar corridors, past elevators, and nurses' stations, and peeked into the ICU. Beeps, bells, and hushed voices came from all directions in this special department. Parents were holding their ill and wounded children's hands, and a newlywed was crying as she watched the slow rise and fall of her groom's chest. There was so much sadness and fear in this room it stifled him. With slightly morbid curiosity, he rounded the corner to the next room to see a beautiful little girl lying in the bed he had once laid in. She slept soundly despite the IVs and bags attached to her frail, little form. Faint traces of blood seeped out from under the large bandage masking her pale, soft forehead. Her dreams were light and golden. She would live to see her scars fade, and he was happy to know this.

Jonathan trudged on again towards his destination. It wasn't essential to complete this task; however, it might speed up the process of bringing her closer. Reaching the large double doorway to the morgue, he waited patiently for the staff to come along and grant him entrance. Not being able to walk through walls or operate a security door was inconvenient but tolerable. He was pleased enough that he could open regular doors and function in the living world at all.

To find her was not a problem for it seemed their fates were bound - maybe, he liked to believe, always had been. Like a pre-arranged marriage, without the aggravation of not being able to choose. A destined relationship, a spiritual connection to achieve a goal needed for their mere existence. He opened the door to where her beautiful body lay cold and alone, drew the white sheet from her head and stopped short. Seeing the residual scowl on her delicate but hard face drew a frown on his. This might be a problem. She had not yet accepted entirely. He pushed his growing sense of impatience aside, and set out to perform the task he felt would be vital. Reaching into the pocket of his faded blue jeans, his hand connected with the amulet with a jolt of cold fire.

Yes, this was right. The chain retrieved, he bent slowly to place it gently around her lovely neck. Positioning the pale amethyst charm in the center of her chest, he closed his eyes briefly, feeling an overwhelming need to hold this cold body but knowing the soul he desired no longer inhabited its human form. Gently placing a kiss on her cold lips, he silently vowed to kiss these lips again. His dark brown hair fell forward, gracing her cheek. Surprised with the intimacy he felt and acted upon, he quickly reminded himself of his purpose He felt immense relief now that this sacrament was finished. He knew it was right, despite the disbelief of the old man. He put things back the way they were, wondering how long it would be before someone realized the corpse had gained a treasure. Softly chuckling, he left the room and the hospital to find the old man once again.

2

First impressions and the warning

I felt a light ticklish wave brush across where my cheek and lips would be. I felt it. It was a marvellous feeling, since I had not experienced any touch at all in so long. Was I asleep? Was it possible to sleep? I quickly forgot the sensation that had snapped me from the unconscious state that I must have been in. I looked down once again. It was day. What day, I didn't know. I was, of course, still at my bus stop. My death spot. Unfamiliar people came and went on the city transit to the required destinations that enveloped their busy lives. It was like watching a movie, all the characters playing their parts according to a well laid-out script that apparently included my death. The crime scene had dissolved, and the area returned to its pre-death state without a hint of the calamity that had taken place such a short time ago. I was erased, cut from the film, no longer needed to enhance the quality of the show, an insignificant ant stamped out without leaving even a smudge of residue behind on this street.

The realization that my family must now be aware of my death sank in. I wished again that I could have been there to see the scene as it had played out in my own home, so long ago. On that day, the police dutifully informed my family that my brother had decided that he, by means of my mother's prescription sedatives, would be better off dying in an alley next to his girlfriend's house rather than enduring life without her.

The tragedy had not torn my family apart, for we had already been distant, non-unified, separate, a functional family facade. My sweet little sister was now left alone with more grief than any 7-year-old should have to bear. Heather had never been much of an involved mother. Having a child had destroyed her acting career, and she held in resentment like a dam. Flipping from man to man, relationship to relationship, she had finally settled down when she met Jacob and had my sister. Although she ceased her frivolous, entertaining, and dynamic lifestyle, she remained intense, remote, and an obstinate alcoholic, which Jacob tolerated with love and acceptance. When Kieran had died, she again transformed. Her grief, regret and remorse brought on the face of a tormented woman, quiet, alcoholic, a newbie but successful real estate agent whose mothering duties and abilities consisted of the occasional housework, meal prep and heartfelt lectures about life and its choices. She did not go to school functions, parent-teacher interviews, ballet recitals, or basketball games. She worked, and she drank.

I remembered when Jacob came to live with us. I had never met him, although he and my mom had been dating for over 6 months. He was funny, shy, and kind, though a bit strange and temperamental, a retired football star who had succumbed to a double knee injury that left him a sad, broken rebel without a cause. Then along came my mom, his new cause. That he loved her, I was sure. That Jacob desired a stepdaughter, and a child of his own, I was not so sure. However, he took on most of the household chores that my mother neglected, began selling encyclopaedias, and spent more time with me than anyone else that had come in and out of our lives before.

During the last few months, I had not spent any time with him or my mom. I had filled my time with Audrey, Nicole, and Willie. My three best friends. We went to the mall, we shopped, we watched movies, tried smoking, and chased boys (well… Willie chased girls) and we stayed out of trouble. We were ordinary, unremarkable teenagers to the rest of the world; to each other we were the world.

Audrey Shaw, 17, was far more beautiful than either Nicole or me. Her natural ash-brown hair and gray-blue eyes made her the envy of most girls, me included. She was perfect, with one exception: the large birthmark that spread like an eagle's wings across her back and right side. This mark alone kept her in our misfit clan instead of upgrading to the elite prep group, even though I knew our friendship would not have suffered one iota if she had decided to become one of the 'model' girls. I had known her since she was picking bogeys out of her nose and wiping them on my shirt. I also knew that her parents were everything that I had ever wanted and that Audrey had a future, a real one.

Nicole would go far in life as well, but not only with her beauty. Although considered pretty by most, her red hair, freckles, and short mousy appearance did not interfere with her using her keen sense and intelligent mind to succeed. She was an awesome friend who was always there when you needed her, without the bogeys.

Willie was different from all of us (not only in that he was a boy). He made me feel lucky and thankful for the family I had. This boy did not have an alcoholic, distant mom. Willie didn't have a mom, and he had a drug-addicted father who took a belt to his remaining son whenever he felt like it. Willie was alone in the world in a sense that most did not understand. He needed us, and we loved him, secretly wondering behind his back what future he would have if his father did not kill him first.

For the first time since I had died, I realized painfully that I would never be hugged by Willie, entertained by Audrey, or humoured by Nicole again. They would go on with their lives without me. I would no longer be a part of the group that had kept me sane for so long. I missed them, each one of them, more than I missed anyone. They had never failed me or left me, though I had just abandoned them. Willie would take it the hardest I was sure. It was his house we would go to for music and video games. It was my house he would sneak into after his father would get a hold of him in the worst way.

It started out that he came to me because I had the most easily accessible windows and parents that would likely not care he was there. (Well at least they didn't accuse us of having sex). Then I became the one he needed to keep rational and sometimes safe. To wipe away the blood, apply the bandages and to understand what it meant to wait for the 'right time' to leave, to move forward. Willie was waiting until he was 18, just 1 month away. He had nearly enough money saved and a plan in mind.

He was leaving Calgary and going to a smaller city or town to work and get his own place. I reminded him that he should also finish grade 12. He would no longer be his father's punching bag.

I had thought many times about going with him, even though I would not be 18 until April 29th. However, I feared that would complicate our friendship and give Willie expectations that I was not willing to fulfill. I had always known Willie had a crush on me; now I wished I had told him that I cared for him deeply, not romantically, but deeply just the same...

I wished, immeasurably, that I could move from this place. I could not make myself believe I was bound here; I could not comprehend that each dead soul was trapped at the last spot they had been alive. Maybe I was a lost soul. I had read books and seen movies where a person had died and could not pass on to Heaven or (god forbid) the other place down below. I now feared that was my fate (apparently dead people are irrational, or maybe it was just me).

Possibly, I was one of them, tortured by things not accomplished, trapped by the inability to let go of my human existence. But that didn't seem right. I knew I was dead, and although I would miss my cherished friends, my little sister, and my parents, I was not tortured, though I was angry that I was gone, confused as to what had happened and scared of what was left to come, but I wanted to move on, I truly did. I was ready to leave and accept my eternal fate, whether it was to be among the angels I so wanted to believe existed or the evil scum of the death world. I wanted something to change.

Slowly, I began to realize that every time I was engulfed by my living memories, subconsciously rolling along the train of thought, that time seemed to go by extremely quickly. I wondered if this was a state of sleep. Taking in my surroundings once again, I realized it was night. The blackness was so deep, so complete, that I could see no stars, no moon, and no glow from the streetlights. To my immense surprise, I could not see the bus stop. Was it just too dark or was it actually gone? It wasn't possible that I was still in the same location; I had to have moved. Fear gripped me in the absence of light. Now I wanted a foot to kick myself for not wanting to see that cursed bus stop (at least there had been a street light). Waiting was all I could do. Pray, maybe, that something would happen to reveal an end to this. Right now, any end would suffice. Feeling the pitch-black darkness around me surprisingly called to my memory the last time I had seen a night almost as black as this.

I was 15 at the time, and I had fallen in and out of love that evening (so dramatic). I was in the middle of my first high-school year. I loved grade 10. Audrey, Nicole, Willie, and I shared many of the same classes. I was getting my learner's licence, and there were loads of cute guys in my new school. The extraordinary boy I met that dark night must have been from a different school, maybe a different city (he was just too hot for my area and its usual specimens). I had been searching desperately for my favourite sweater, which I had left in the gym bleachers. A group of boys and their parents started coming into the gym for the basketball tournament that weekend. I was scheduled to meet up with Audrey and Willie by seven p.m. to go watch the latest blockbuster (*Twilight*, of course - I'm so in love with vampires), at the theatre. I was going to be late. Staying after school as part of my punishment for not handing in two of my English assignments (I honestly lost them: computer crash) gave me the opportunity to sneak in the gym before the game began.

Finding my sweater hanging by a sleeve high up on one of the open bleacher steps, I grabbed it quickly and started to jump down the bleachers and run to the bus to make the movie. I realized in mid-jump that this was going to end badly for me. And it did....for a moment. An untied pink lace of my new multi-coloured Converse got lodged between the benches. I reached the end of the lace and fell hard to the bottom bench with my left leg still suspended up two steps. Pain shot through my right hip and ribs as I felt the heat rush to my usually pale cheeks. Oh, how embarrassing to wipe out in front of so many tall, undoubtedly good-looking future NBA stars! That was just my luck. My grandma also used to tell me if I didn't have bad luck I would have no luck at all.

So, there I was, lying upside-down on the bleachers while being snickered and outright laughed at by many now-not-so-potential boyfriends. Fighting back the tears, gritting my teeth, I attempted to get up and quickly un-attach my lace, or remove my shoe, chew off my leg, whatever it took to be freed from this embarrassment. My hand was just inches away from my right foot when this pale, gentle, smooth hand grabbed mine. I felt a buzz run up my arm like the faint electrical shock you get from touching a vibrating seat cushion. Startled, I looked up into the most enigmatic face I had ever seen. Well, adorable, handsome, gorgeous, amazing, and earth-shattering were all words I could use to describe him. His unusually long-for-the-times dark-brown hair was falling lightly forward on his forehead, just above the palest blue eyes I had ever seen. Even Audrey couldn't compare to the magnificence in this face. He couldn't be real. But there was no doubt I was in love.

My saviour wasn't smiling; his full pink lips formed more of a grimace. With a quick, fluid movement, he detached my lace and placed his strong arm around my tender waist to lift me to my feet. I stood up on the bottom bench, staring down at him, jaw dropped, eyes wide. I must have looked like a total dork. (I usually wasn't this useless in front of boys.) He still didn't smile when his luscious lips parted to speak. Expecting a voice like music, I was shocked when this deep, husky voice emanated from this dream boy.

"Are you all right?" he asked, slowly forming each word carefully as if the English language were new to him, though there was no hint of an accent.

I couldn't speak. The shock of his touch, his perfect face, physique, and unexpected voice had me speechless like a deaf mute. My embarrassment worsening, I nodded my head 'yes' and waited for him to speak again. He didn't. He stared at me, then brushed his delicate fingers against my cheek. The touch was so sudden and so unexpected I forgot to breathe. I couldn't understand why this breathtaking boy would help me, or touch me with so much sensitivity. If the pain of falling weren't so miserable, I would undoubtedly have thought I was dreaming, or worse, unconscious. Not that I was hard on the eyes, unpopular or had a giant pimple on the end of my nose, but I was still flabbergasted that this so obviously perfect boy would come to my rescue.

The crowd was now advancing on the bleachers and gathering around us, when he suddenly motioned for me to follow him. He led the way out to the large, full parking lot and stopped still to stare up at the sky. It was dark: a new moon and no stars in sight. I thought this to be just an odd addition to a supremely odd situation (and frankly, an odd boy, not matter how scrumptious he looked). Neither of us spoke as we walked to the bus stop. He must be taking the bus too, I thought, but then why did he just show up in the gym in a crowd and then go to the bus stop with me? I couldn't understand, and I was starting to think I had banged my head, and this was all a figment of my swelling imagination.

Looking quickly at my watch, I realized that I had missed the first bus by minutes. My best friends were probably already in the theatre settling in with their heavily buttered popcorn, liquorice, and drinks, ready to watch two of the most beautiful people (well one vampire) on Earth fall in love. I had wanted to see this movie more than anything since reading the books. The story had become an escape from my dreary life when I wasn't with my crew. For lack of my own love life, I was nearly completely satisfied in enjoying a fictional one (and it didn't hurt that both Edward and Bella were incredibly intense people).

I decided to wait anyway for the next bus and be better-late-than-never for the movie. Maybe this darling boy would be going too, and we could walk in together holding hands with a carnal passion that would put even the actors on the screen to shame. Blushing again, this time at my ridiculous thoughts, I glanced over to my unlikely saviour. He was frowning again, in almost a pout. His intense eyes were flitting everywhere, unseeing, as if he were going through immense mental anguish. I saw him unclench his fists as he turned towards me once again.

"Well, you're okay now so I'll go...." he said tentatively.

He stopped speaking as quickly as he had started, although it appeared he had more to say. I waited to see if he would continue, but he didn't. I didn't want him to go. I didn't know why exactly I was so drawn to this strange boy. It couldn't just be his astonishingly luscious looks (was I that shallow?), or the fact that he swept into the gym like a knight in shining armor to rescue me, a forever-damsel-in-distress. Whatever the reason, I felt immediately panicky when I saw him start to turn in the opposite direction of the school. Strange, I thought, when realizing I was pretty sure he had entered the gym with the large group of player's parents, and supporters, and was now walking away just a few minutes into the first game. Was I wrong? Was he not from the 'out of town' team? Could he be from here, this city?

"I can't just stay here and make sure you get on the bus okay, you'll be fine," he said suddenly.

A confused feeling of indignation came over me. He was being rude! Acting like I was the one who had asked him to come and save me. All he did was detach my shoelace - I still hit the bench, and I would still have bruises from my fall. I had taken the bus three or four times a day for as long as I could remember, so he had no right to assume I would not be able to get on a bus safely. (Wow, I can be so childish.)

"Of course, I will be fine, I am not worried about getting to the movie theatre, I was just....ummm...going to say thank you."

I lost the nerve to tell him off for his baffling tactlessness, though I immediately decided I was not going to give him the satisfaction of my asking him to stay and chat or even possibly of going to the movie with me. I was still stung from his apparent lack of confidence in my ability to take care of myself.

"Yeah, no problem, it was just a shoelace," he said, appearing to read my mind.

"I am sure you will get to the theatre to meet up with your friends and have a fabulous night," he sneered with a look of impatience that suggested that he still had unspoken words or an unfinished agenda. I didn't even clue in that he knew, somehow, what my plans were for the evening.

"I know l will have a lovely night, I can't wait to see this movie, I have wanted to for forever, and Audrey, that's one of my best friends, she is going to be furious that I'm late, I am sure they will still let me in if there are seats left, which might be a problem cause it is a hugely popular movie...Twilight, you know, Vampire loves silly human girl," I rambled stupidly, quickly realizing that I was giving away too much information. This guy, although adorable, could be a creep, a stalker, or some other crazy person. But for some reason I still had this overwhelming need to keep him standing here.

"You'll get in. Well, goodbye."

He turned again, this time making it all the way to the edge of the bus stop before I yelled.

"Hey, I don't even know your name. Could you just stay and chat until the bus comes?" I felt my cheeks reddening at my forthrightness but did not care enough to let the pleading smile I offered him leave my face.

"I have to leave now, I have stayed too long."

Shocked at the anger and surprising sadness in his voice, I recoiled slightly and ceased smiling.

"Oh well, fine, you go on now and do what you got to do."

I knew I was being impolite, but I didn't care. I was miffed at his reaction, thinking that if he didn't want to at least tell me his name he should not have helped me in the first place.

"I can't be here anymore, Anna."

I couldn't speak right away, although it seemed that he was waiting for me to say something, desperately searching my face for evidence that he wasn't wrong in helping me. I surely couldn't understand why he would want to hear what I was thinking, knowing that it was not polite. I started towards him, staring at his pained blue eyes, his slightly messed thick hair, noticing for the first time that he was wearing dress pants and a dark blue button-up shirt. He was also wearing black dress shoes. Odd apparel for a basketball game.

"I 'm sorry. I am sure you must have somewhere important to go, being dressed so nicely. Hopefully you will be fortunate enough not to run into me or any other silly girls again tonight."

Despite my anger and hurt feelings, I was stunned to realize that I was still hoping that my words and underlying guilt trip would elicit some sort of information from him about where he was from or something to indicate that he didn't think I was just a clumsy, accident prone idiot who was demanding his attention.

"The next time we meet, it might not be under such easily fixable circumstances." With this presumptuous and perplexing statement, he turned once again and walked quickly away. I shut my eyes in a feeble attempt to squelch the tears threatening to spill over, and when I opened them again he was gone. Not just far down the sidewalk, not driving away in a car, but gone. Hearing the bus pull up with a squeak of the brakes, I grabbed my purse and got on.

The first thought I had after replaying the whole scene in my head was that he was so lovely, and I had been so belligerent, and he was so... ohm something else not polite and too polite at the same time. The second thing I realized, with a puzzling jolt, was that he had said my name. He had called me Anna, a name no one had called me since childhood. (I must have been hearing things.) The unlikely meeting, mystifying verbal exchange, and unexplained moments were put aside when I arrived 20 minutes late for the movie, but the image of his face remained throughout the night and for many days to come.

* * * * *

It wasn't a positive reaction he received when Jonathan came back to the middle world after seeing Adrianna at her school that day, and he had spent many days after brooding over her disdain. But most of it was long forgotten until he met up with his old friend.

"What were you thinking, Jonathan? Do you have any idea what you could have done by watching her so closely for so long, meeting her while she was alive, helping her in ways she shouldn't have been helped, and placing that amulet on her body?" the woman said scornfully. They met in their usual place. Jonathan was saddened that the first words she had to say to him were ones of disappointment. He hadn't seen her in quite some time. However, most were used to her long absences.

The gardens had always inhabited an exceptional place in his heart. Walking through the tree-lined trails, the branches shading the walk to almost dimness, dropping in places so low you had to duck, flowers of every color, shape, and scent, lying in beds bordered by white and gray stones that could have come from heaven. The tiny stream carrying leaves along the bends and twists of the 5-acre gardens brought a fresh, moist feel to the fragrant air. He longed to sit on the old cast-iron park benches, feeling the warm breeze, hearing the morning birds sing and the pigeons coo. It was a pleasure to listen to the children laugh as they waded in the stream, fed the birds, or threw a ball to their dog. It was his favourite place and the place he had first seen Adrianna.

The woman waited patiently for his retort. She was angry, this he knew by her rigid posture, thin-pursed lips, and the tone of her voice. She sat on the bench beside him, her legs crossed under her silky mauve dress. Her white-blonde hair caught the rays of sunlight like a solar panel and shone. Her blue eyes were closed as if she were shutting down to gain a bit of control. She also understood. This he knew from the softness in her touch when she held his hand. She had to reprimand him even though she would have done the same thing, and that she had done the same so many years ago. However, for her it had not gone so well; for him, it would, he was sure.

"Jeez, Juniper, you know I had to give her the stone. You know the power it can contribute to the right one, and Adrianna is the right one, I know she is. I have played by most of the rules. She has only seen me the once and will not remember seeing me, not that it matters. I have watched from a distance as I have had to do. It has been hard, and you know how that feels."

He tried to appeal to her own experience of waiting for the one she loved to die and join her. He explained that he had acted by the book. She did not know about the visits to watch Adrianna sleep, or the times he had saved her life when he couldn't bear to see her life end that way, or the gifts he had left for her.

"Jonathan, if she is to be as valuable as you and some others believe, then you had better make sure you do not do anything to prevent her destiny from taking place to the full potential. I do not share your beliefs, and neither does Shadow. He has been quite lenient with you, allowing you to have more of a physical presence than you should have. Don't push his kindness or my understanding." Bitterness crept into her voice at the end, and Jonathan knew she was remembering her past. The pain she had suffered was clear, as was her desire to see none go through what she must bear through eternity.

"Holy, I hear your warning, as well as Shadow's. I am not ignorant of the danger here or the potential for loss," he added for her benefit.

"But I know I am doing the right thing; she will come around and gain strength. The amethyst will speed up the process of reckoning, and she will fulfill her purpose for all of us." He hoped he had put her fears to rest or at least settled her mind. The lines of worry did not soften from her beautiful pale face; however, he knew she would eventually see that he was right. He did not expect the words she said next.

"Can you be sure that you are focused on her presumed latent powers, and not on your affection for her?" She did not wait for a response. Her disappearance would have sent a whirl through the mind of anyone who had seen her sitting there; however, he had been subjected to her quick departures, always with the last word, for as long as he could remember, and that was a long time.

Allowing the question to penetrate his shield would not prove anything, as his own thoughts damaged his focus enough. He knew that having any feeling for Adrianna, other than the desire to help her to become what she was meant to be, was dangerous. It was not allowed for legitimate reasons.

There were many rules and those rules needed to be followed. He despised the rules though he begrudgingly recognized their importance; he also knew the destruction that breaking the rules could cause. Souls lost for eternity. Like his mother. He couldn't bear to bring those memories to the surface. They had been buried for so long. He dismissed Juniper's words, believing they came from her own loss, not permitting the thought that she might be right.

Staying on the bench, he closed his eyes to draw the vision of Adrianna's face. The feeling overtook him entirely, the need to protect, the strong desire to see her be what she was meant to be, the satisfaction he felt at knowing he would be a part of saving his world as he and so many others knew it. The immense longing to hold her, touch her face, caress her lips with his, hear her gentle voice and gaze into her serene brown eyes was forcibly denied and repressed for her own safety as well as his.

He had a job to do. And he would do it. He would show Shadow and Juniper that this was his quest and his alone. Adrianna was bound to him and him to her. There was a purpose for this. He would be the deliverer, the one who saw her for how exceptional she was, recognized what power she possessed and how that could be used to save them all. Anna would see it and so would everyone else. For millennia to come, her name would mean freedom, connection, and heaven. And he would be the reason it all came to be.

3

Funeral and fading

Still dark. Why is it still dark? I want to see something. I wished for my bus stop. If I wished it away, could I not wish it back? Trying to concentrate on my body that was no more, I realized that I could feel a slight pressure where my neck would be, though I could bring no hand to touch it. The feeling was as startling as the earlier touch on my non-existent cheek and lips. Any feeling of touch was surprising, bewildering and oddly comforting.

Is this truly death? I would not have imagined this. I would have thought I would just be gone, erased, no soul and no ability to think or feel. Or if my mom was right in her lifelong beliefs, should I not be sipping margaritas with George Washington or discussing the Civil War with Lincoln? This surely was not heaven, but was it hell? Had I done something in my life that could have kept me from the pearly gates and forced me into oblivion?

My mind briefly wandered back to days in my past: the arguments with my mom, the day I dumped a bottle of her 'happy juice' down the sink and she had grounded me from my friends for a month. Or the time I ran into Steph on my bike and she needed six stitches to minimize the inevitable scar on her knee. There were times I had lied to my friends, not wanting to go to a particular movie or having to candy-coat a few words about new outfits. Skipping school a few times and only cheating once on a small math quiz surely wasn't enough to bind me to the darkness I was surrounded by. I had always been there to help my friends; didn't that count for something at the end of all ends? (I wasn't Mother Teresa, but I was still a loyal friend.)

Willie said once, a few months ago, that I was an angel. He had been in my room for a few hours, and it was nearing 5 am. We had cleaned his latest wounds, and I had applied butterfly bandages to a gash that should have had stitches. Willie had cried on my shoulder until we started to see the first sign of the morning sun touching the horizon. I promised him that I would do anything I could do, any time I could, to help him. I also promised I would always be there for him before and after he left his painful jail with his father. He called me his angel, and said that he would not know what to do without me.

My grandma told me too that I looked like an angel, the day before she died of lung cancer. I had put on my prettiest peach dress with a white satin sash, to read to her as I did every Sunday. My heart always belonged to those I loved. I gave it easily and expected nothing; maybe I should have asked that it be recognized when I died. Oh Gran, your words of wisdom…

Seeing Gran's face in my mind made me want to cry. I missed her so. It was 4 years since we had spoken. Sitting next to her in the hospital those last two tear-filled months was the hardest thing I had ever done. Singing songs, reading stories, wearing her favourite knitted sweater, which she had made me years before, was all I could do to try to reach her as she lay fading away in the hospital bed. The flowers I had bought with my babysitting money every week still sat in their pretty glass vases on the windowsill, some without a petal or leaf left to show how beautiful they had once been. Gran would mumble a lot; a few times, it was my name. But then she died. And I was left alone with my complicated family and my sorrows. Maybe, if I could find a way out of this pitch blackness, I could look for her. I wonder if it would be like looking through a mall or just an open room filled with people who once were somebody, had lives, homes, children, parents, and friends.

Like a jolt of surprise, I saw to my right a light slowly spreading towards me. Out of instinct, I willed my body towards it, not believing this would work, but to my surprise it did. I began to glide slowly in the direction of the light, but the light did not seem to be spreading my way anymore. It seemed that as I glided, the light stayed with me and we glided together. Patience was apparently a virtue that helped as much after death as in life.

Willing myself to stop moving made the light approach again. Staying mentally still, I waited for the light to reach me as I stared intently, searching for any forms in the light. From somewhere in the distance, I thought I could hear music. I began to focus more on the mounting sound and willed my mind's eye to stop seeing the light. Again, I heard the faint hum of what sounded like an organ, as from a music hall or church although the song was indiscernible. Fear was stabbing at my mind, though my curiosity and relief at the change in scenery overcame any thoughts of danger. (Honestly, I was dead, what could possibly happen to me?)

Shapes were starting to form in the distance, and I was sure I could see many people gathered around a table. As the song became more identifiable, I quickly recognized it as one of my childhood favourites. With complete and utter amazement, I realized that I was in a church, listening to Amazing Grace and watching my family and friends console each other over their startling loss....

My mother stood beside Jacob. They were showing two women I didn't know pictures of me as a child: one where I appeared to be riding my bike, and another of the only ballet class I ever attended. There were other pictures scattered on a lace-covered table surrounded by lit white candles. Steph was standing next to my mom, holding the pink bunny that I had bought her for her birthday the year before. She had sworn up and down that she was too old for a stuffed animal, but she slept with it each night afterwards. There were many people I did not recognize, and many I did.

I then noticed Nicole, standing with her mom and older brother Galen. She was crying and looked as if she had been for days. She always was acutely sensitive. I had to chuckle when I noticed she was wearing the exact dress her mother had picked up for her at a 50% off sale, saying it would be proper for her next prom; Nicole and I had agreed that it could only be decently worn at a funeral.

I could see many of my classmates from this year and a few from other years before. I guess I've had many friends. I was surprised to see that even Shianne Parker was here, looking rather uncomfortable, snapping her gum impatiently. Then I saw why she had attended: my ex-boyfriend Calum (her current love) was here chatting solemnly with two other guys from my Math class. Many had tears on their cheeks and looks of disbelief. I guess it would be terribly hard to imagine my being dead. I was young, smart, popular, and a decent student. I had my whole life ahead of me, and great aspirations. Everyone who knew me knew that I was planning to be a nurse, preferably a paediatric nurse. All my close friends knew of my love for children, as I had babysat for my spending money since I was 11 years old. Helping people was what I wanted for my life's work. My parents had a good-sized college tuition fund set aside for me, which I added to on my own if I had money left over from my shopping sprees and weekend adventures. I had also been saving for a car. So many plans and so many dreams.

Surveying the large group of mourners, I located Audrey; as I might have expected, she was crying and sitting quietly by herself. In her lap, clutched in her soft, small hands, was my favourite purple sweater, which I had lent her 2 weeks before. It always looked better on her than me. I hope she keeps it. Wanting to reach out and hug her, Steph, and Nicole with all my might was too much for me to bear. I started feeling pain where my chest would be. The feeling took me by surprise, as I had only felt that pressure on my neck and the faint brush on my face since I left my body. I remembered the pain well: it was the pain of loss, the pain of loneliness and the pain of regret.

I wanted to tell every person in the room (well, the ones who were there because they cared) that it would be all right. That I was still here, somewhere. But I couldn't. I couldn't do anything at all but watch helplessly as my loved ones reminisced, hugged, and cried. The pain growing inside me seemed to fill the void where my body should be. It was everywhere. I would not try to turn away or shut my eyes to float away mentally until I found Willie. I worried most about him.

Glancing around again, I saw my mom kneel over with grief and Steph start bawling. I wanted to scream out that I was here — please look at me, see me, I see you. Searing heat travelled through where my skin would be and I had the strange feeling of being electrocuted. I asked for this. I had this sick curiosity about what it would be like to see the aftermath of my death, not cognizant that it would include my funeral, not thinking of who would come, who would cry, what it would look like. Now I was seeing it, and it was worse than dying all over again. I needed to find Willie fast and then pray that I could leave this place and never again be witness to so much pain for so many people.

Searching through the sea of now-faceless people, desperately seeking the one face I needed to see, I began to worry that he had not come. Maybe his father had trapped him in the house somehow; it had happened before. Maybe it was too much for him to come and see my blank face in the beautifully laid-out coffin. Just as I had resigned myself to not seeing Willie, there he was, sitting in the front row of pews closest to my open casket. In his hands was one single white-and-yellow lily. Only Willie had ever brought me my special flower. I was almost positive he was the only one who knew it was my favourite. His handsome face was drawn and he had deep-purple circles under his gray-green eyes. It appeared that his light-brown, shoulder-length hair had not been washed or combed since the last time I saw him. Still, he looked so gallant in his grey pin-striped suit, reserved only for special functions. His six-foot-two frame was slumped, and I was pretty sure I could see a new bruise shadowing his left cheekbone. He still wore the hand-decorated cast on his left arm, where his father had broken his ulna a few weeks before.

If I thought the pain was terrible before, it was nothing compared to the sensation I was now feeling. If I had been standing and alive, I would have dropped to the ground now and begged for death. How I wanted to touch his face. The love I felt for Willie was not a love for a husband or a life partner, it was a love for a soul-mate of friendship, one that could never be replaced or tarnished by time, lovers, or even death.

Willie was my best friend, and I missed him so much. I needed him now as he had needed me so many times before. I needed his carefree way of putting extremes into humorous perspective, his ability to make light of the darkest situations, his solid comfort when I couldn't deal with my mom, losing my brother, or the trivial pains of being dumped or rejected. Willie was always there for me as I was for him. Now, when we both needed each other the most, we could not touch, not comfort, not bandage the injury, not laugh away the fear. I could only stare at his broken body and his tortured face. He could only hold my lily and gaze at my dead, cold, soulless body.

Willie got up slowly. He seemed oblivious to his surroundings: the people mourning, the white candles lighting each square foot and the sad, gentle music always playing in the background. As slow as his walk, he placed the delicate flower on my chest just below the amethyst pendant my body now wore. (What is up with that? Must have been a post-death gift.) The pendent was remarkably pretty and gave me a new sensation, one of longing and anticipation. Ignoring the unexplainable emotion, I turned my attention back to see a single tear roll down Willie's cheek and onto mine as he lightly kissed my lips. I saw his lips move to words I could not hear. I assumed he was saying goodbye. The agony took over again, as my surroundings started to fade. I desperately clung to the painful scene, to seek out my family again. Steph was now sitting on the floor, chewing her fingernails, with tears still streaming down her now rather flushed porcelain-perfect face. The vision faded just as the same woman I had seen at my death looked curiously up at me from a crowd by the organ.

My funeral vision washed out much as it had appeared, in a gradual wave, this time away from me. Blurred faces, music and distraught murmurs were there, then gone. Being left without a visual for the anguish I felt was easier in a way, although I longed to hold on to the sight of my loved ones. There was a residual light left long after anything discernible had gone. I was glad of this. With each passing moment, the light increased just enough to notice. I began to believe hopefully that this meant I would not be in the dark again. I could still feel the light weight on my neck. I concentrated on the only physical feeling I had besides my grief. The heaviness was now all around where my neck would have been, the heaviest part being about two inches below my throat. My thoughts flickered back to the last seconds of my funeral, when I had gazed timidly into my casket and seen the unusual new necklace and charm my dead body now wore. The chain had been silver and sparkled in the candlelight, the charm had been the palest amethyst. It was beautiful, and somehow, I was wearing it. I wondered who had bought it and placed it around my neck. It did not seem like anything that my family or friends would buy. My birthday was in April, which was represented by a diamond. I was sure that the amethyst was for February. Wondering if I would ever know where it had come from brought to mind a buried memory from a Saturday long ago.

The number 18 bus had sped from my stop just seconds before I reached it. Cursing under my breath, I settled onto a seat in the Plexiglas enclosure for the 30-minute wait. A young mother, holding hands with a small child in a red raincoat and pushing a plastic-covered stroller toting two more little ones inside, walked towards me. I felt sympathy for the rain-soaked woman and moved out of the way for her to bring in the stroller. I could hear the furious cry of a baby and my sympathy grew. The walking child stood outside in the drizzle holding a blue balloon. I smiled at the woman as she lifted one of her twin girls from the stroller and started feeding her.

I was slightly embarrassed at first at her apparent lack of modesty, but realizing that this was the rather typical reaction of a teenage girl, not of a would-be nurse, I quickly glanced and nodded to the woman again, showing her and myself that I was more comfortable with the open breastfeeding than I had initially felt.

The balloon-child was playing in the light rain, in front of the bus stop, giving peals of laughter as the balloon bounced up in the air each time it hit the ground. The force of impending doom hit me just as a gust of wind from the north tore the balloon from the little tot's hands. I lunged towards the child, his wet coat slipping from my grasp, the mother just looking up from her feeding baby to hear the screech of tires on the slick pavement. The driver of the black Ford Ranger had apparently seen the child just as he jumped from the curb towards his floating toy. However, the road was wet, and the truck could not stop. As I ran, I envisioned the child being taken from this world under the truck because of a floating piece of rubber. Seconds before the death (or severe pain and injury), I finally caught a hold of his little arm. Yanking with all my might, ignoring the danger of tearing his arm off, somehow, I tumbled back onto the curb, the little boy landing on top of me. How we landed safely, I didn't know. I had been running forward and was going to grab the child and keep running forward in the hope that the looming truck would miss us both. My head rang with pain from hitting the cement. Surely, I would be black and blue after this.

The silence was deafening until a scream came from the bus stop. The driver also shouted as he came to us. I hugged the small frame as I got us both standing up. The man touched my arm apprehensively as the mother appeared in front of me to tenderly grab onto her son. The whole situation took just moments to begin, and moments to be resolved. There were tear-filled thank-you's, handshakes, more tears, and then a whopping write-up in the local newspaper. My parents didn't know I had it in me; frankly, neither did I. I sure wanted to believe it was my destiny to save people in peril. It was a brilliant destiny, despite the pain I was in for many days after saving little Keifer. I received the second Christmas card this past holiday from Sabrina, little Keifer, Zoe and Zayna, thanking me once again and wishing us all a safe and merry Christmas. Little did they or I know it would be my last one alive.

"That's right, Adrianna ... remember." The rough voice came out of nowhere and everywhere, echoing around me, shaking my internal stability like an earthquake (not that I was that stable to begin with). I hadn't expected to hear voices, especially one saying my name. Maybe there was a god and my mother wasn't wrong for dragging me to Sunday school for three years. Since I died, I hadn't given much thought to 'God'. I had never actually decided if I was a 'believer'. I had attended Sunday school and the occasional sermon, but it wasn't until my brother committed suicide that I questioned my ignorance. The pastor who spoke at his funeral never mentioned that he had taken his own life; he also didn't say that Kieran would be accepted into God's kingdom as all God's children were. At school, the verdict was confirmed. Bobby Sprint, an old nemesis of mine, told me every day until he grew bored of my weeping and torment that my brother would surely go to Hell because that's where all suicides went. Heaven didn't have room for a coward and cop-out who swallowed a bottle of pills over his girlfriend's finding a new love.

So, my acquiescence to my mother's occasional preaching was based on many mixed feelings. And now, when the opportunity might arise for me to prove or disprove Heather's drunken teaching, I had a little more than a slight curiosity. Finding out how to make the light materialize and stay was much more beneficial to me.

The moment it dawned on me that satisfying my religious queries was in distant second-place to obtaining more visual acuity, I saw that I had already done this. The brightness around me had doubled, at least. Although I was not in the funeral home, or anywhere I knew, I could see a floor beneath me, a gleaming white floor. There did not seem to be any walls or boundaries of any kind, nor were there any furnishings, but there was a remarkably clean-looking floor. I still had no feet to walk upon it, though. Would I ever?

Saying a small prayer from childhood, I begged for the voice I had heard to speak again. Although I was almost certain it was not a voice I recognized, it was still a voice, and not the depressed one in my head that kept talking about how horrible it was to have been at the very funeral I had wished to be present at (from an outside perspective, of course). Not understanding what the voice had meant by my being right to remember made it even more desirable to hear it again. I did enjoy remembering saving little Keifer, he had turned out to be a sweet, darling little boy. Babysitting him on the few occasions I could was a joy, and not one I would easily forget. But why was my remembering anything at all relevant to my mystery speaker?

I had hundreds, maybe thousands of memories from my short 17 years. Some were wonderful and others best left forgotten. Willie's tortured face was one new memory I could not seem to shed. There was something about his pose, his wilting shoulders on his usually strong frame, and the desolate look in his intense eyes. I felt the return of the ache from my out-of-body funeral experience. The illumination around me flickered and dimmed as if feeling my thoughts. It seemed impossible not to be concerned about Willie when I had so willingly worried about him for so long. I felt as if there were something I was missing, some connection that needed to be made. It puzzled and scared me that I could be ignorant of a piece of information that might help my best friend; or maybe I was confused, and it wasn't about Willie.

"Stop thinking about him. You can't help him now."

Startled and delighted at once that the voice had returned, I also felt a minor irritation that this voice thought it could tell me who I should and shouldn't think about, and who I could and couldn't help. I would do anything for Willie, even after my death.

"Let it go. You have more urgent things to achieve," The voice commanded.

Hearing the vehemence in the words made me cringe, leaving the annoyance behind. Still wanting to hear words of any form, other than those in my own head, was enough for me to give in and decide to follow the advice given, if just for now. I pushed the disturbing thoughts of Willie to the recesses of my mind and began to concentrate on more positive thoughts and memories. As I remembered days spent laughing with my sister, endless shopping trips with Audrey and Nicole, and the day I received an award for creative writing from my English teacher, I absently noticed that the light around me was gaining in intensity. I could now see blurred lines of where walls might be in a room that would prove to be extremely large.

Feeling elation at this increase in visibility, I carried my thoughts onward to include more intimate aspects in my short life. Kissing my mother's cheek when she, after being hospitalized for drinking, had finally woken up and promised that she would never again allow herself to be out of control. Comforting Nicole when her dog, after 16 years of life, died in her arms. Willie bringing me a bouquet of white lilies for my 16th birthday at school and presenting them to me in the lunch room, in front of over 800 students. I was seriously embarrassed, but the tears I shed were tears of pleasure.

Without knowing how it crept in, the recollection of the evening I had met the most beautiful boy, came to surface. His passionate eyes had looked at me the way no others had. It felt as if he had known me all my life and decided to pop in to my world and make his presence known. Honestly, before my death, I hadn't thought of this occurrence in quite some time and was surprised that it had resurfaced, twice, after my death.

To test a theory, I had brewing, I allowed my train of thought to travel to my last image of Steph sitting alone on the stone floor. The light dimmed again. My theory was confirmed. I worried about how to keep my thoughts from travelling down the deep, dark path they were so carelessly driven to. The second theory I had was that the mysterious voice would only appear when my thoughts were distressing and produced the suffering I had experienced earlier, particularly when they pertained to Willie. Where was the happy medium, where I could think happy thoughts and still draw on the only sound I had heard since my memorial service?

* * * * *

Jonathan sat at the large table covered in books, paper, and a clock. The room was small and dark, with mahogany paneling and black furniture. More books lined the wall, the titles hidden in the shadows, the only illumination in the room coming from the crack in the green velvet curtains blanketing the solitary window on the far wall. A small candle burned on the right corner of the old colonial desk. Even before seeing the old man enter the room, Jonathan felt his presence. The vibes emanating from Shadow were not of an agreeable nature. He knew that neither Shadow nor the Council would be happy with the small rule-breaking he had been indulging. If Jonathan hadn't been so sure it was all for the greater good, he would have followed all the regulations as he usually tried to do. He prepared himself for the imminent lecture and closed his eyes to summon the image of Adrianna's face to give him the strength and audacity needed to explain his deviousness to the old man.

Jonathan turned his chair to address his friend. Gazing at the vision before him, he stopped. Looking at Shadow's body and realizing he could actually see the wall through him, he stared incredulously.

"Holy smucks, Shadow, what's happening to you?"

"I came to tell you that the council wants to gather. There have been a few....alterations...and um...situations occurring. Can you break away from your mind-flogging long enough to bless me with your presence following the meeting?" His rhetorical question hung in the air as the scorn in his words settled like dust on the desk. Jonathan knew that he had only avoided a blasting from Shadow because of the visible beginning of the end of all they knew. He wondered if the Council members who had all been here for centuries longer than he had, would be as transparent as his friend.

He knew this couldn't be good. He had been warned for many years of the breakdown in the solidarity of their middle world, but he had not been witness to any changes thus far. It had been foreseen, though, that there was a disturbance in the structure of their world, that some of the old ones might start to disappear. That was why she was so necessary, now more than ever. Seeing Shadow become transparent was more than enough to shake his assurance in his position. He worried that she would not be ready in time. Frowned upon as his methods were, he knew he must proceed. Even the fact that the old man had not delivered his reproachful speech showed that the situation was seriously ominous.

Finding out who Adrianna was, and how unique she would be, had been ironically accidental. It was just before one of the elders had decided to move forward. They had been sitting in Jonathan's favourite park, though not on the bench where he usually sat, when two dark-haired girls came over to the adjacent swing set. The elder, by the name of Hunter, hadn't been paying attention until Jonathan made an absent comment on the likeness of the two sisters despite their difference in years. Hunter looked up and started to laugh hysterically. Confused, Jonathan stared at the old man quizzically. Through shorter bouts of chuckles, the elder told Jonathan that it was quite funny that he could show him one of the special ones who were supposed to protect their very existence; however, the old man was so vexed about the prophecy that he was moving forward regardless. Jonathan was astounded and devilishly curious to see how this bouncy little girl (who would obviously grow up to be breathtaking) could be regarded so highly. Without an explanation of either point, the old man made one last comment before he disappeared, never to return.

"And to think no one was supposed to tell you about her, let alone allow you to stumble innocently within feet of her. Goodbye, my young friend; maybe we will meet again."

Later, reading in some of the rare literature that he found in the small library, Jonathan could gain an insight into some of Hunter's unusual words and use his new-found knowledge to plan a way to secure his world. Seeing then, and many times afterwards, the little girl with long, dark, wavy hair, who would be worth more than anyone conceived, prompted feelings that he hadn't expected. She was adorable, stubborn, and altogether too kind and devoted to lost causes. Watching her grow from a gangly preteen to the now all-out stunning young woman she had become was the highlight of Jonathan's daily life. Waiting for her to die was the hardest.

He had seen the difficult childhood. With her mother's devotion to her own life and not so much the lives of her children, Adrianna was left to care for her sister and find her own way the world. She took in birds with broken wings, and later set them free. She said pleasant things to people who had been rude and even cruel to her. There seemed to be no end to the empathy and sympathy she felt for so many dubious people. Jonathan wanted to comfort her each time she found her mother drunk, or gone for days at a time. On the other hand, he had wanted to scold her for wearing her heart on her sleeve and making it so easy for anyone to hurt her. Thinking of his own past, so long ago, shuddering at the differences, he knew he could never be like her. Would he ever be worthy of her? Quickly putting the intimate thoughts aside, he knew it was time to check in on her again.

Adrianna was where he had left her, alone and afraid. He hated this part but knew it was just one of the many steps that had to be taken. She had heeded his harshly delivered advice and kept her thoughts on a more positive note. He was sorry he had missed the memories that had filled her room with light and now a bit of furniture. He knew she must be furious to have a couch to sit on but no body to use for the comfort. It couldn't be much longer. The autopsy was complete, the funeral was over, the burial to follow. Jonathan had been unable to check in to see if it had already happened, though he expected it was all done. The Adrianna that had charmed the living world now only existed in pictures and memories.

Jonathan was slightly worried about the sequence her progression had taken. She had come so far already; it was unusual for the form not to be full when the blackness had lifted. He had pushed her hard though, and the amulet was working its power as well. It was his interference that had changed the usual course. He must be patient and guide her more gently.

Patience would be difficult to maintain. Shadow was fading, the council was meeting, and he had no time left. He had too many conflicting issues: wanting to see her, wanting to talk to her, wanting to help her become and be fulfilled, all in time to save everyone, including her.

4

Willie, Valerie and Shadow

The chair materialized first. I had been staring at the empty location when it was suddenly filled. It was just like a fader on a set of digital pictures: it wasn't there, and then it was. The light around me became equal to what you would expect in an office building, bright and intense. I was wondering how I could sit on the comfy-looking seat when the table appeared out of nowhere, right next to it. What I needed a table and chair for I could not imagine, as I had no body. I felt frustration and despair. Then I felt guilt. I had wanted to leave my home and my family, so I left, only to die at the bus stop. I wanted to leave the bus stop and I received nothing but black all around me, smothering me. I wanted to see my funeral to satisfy some sick curiosity, and I witnessed pain and heartbreak, and I felt like dying all over again. I now wished for light and things to occupy the space around me, and I couldn't be satisfied in receiving it. I had to have my body back, or anybody with which to enjoy the apparitions.

"Be patient, Anna: it will come in time, faster than you think. You have every right to want, to need and to desire. Don't despair, don't worry; time will bring glorious things, and time will allow you to give in return."

The words came from the same voice but not in the same tone. These words were kind, soothing, reassuring, and just what I wanted. Thank you, oh thank you. I knew I was not alone now, not trapped in an empty oblivion. Someone was waiting for me and wanted me to be calm, to think happy thoughts. Someone knew things would be better soon, and I would be able to progress from where and what I was. I could hope again. What could I possibly give in return? It was hard to imagine being anything but an unstable nuisance to anyone over here (wherever here was). It's not as if I could lend a hand, or give a hug. (Oh my, was I honestly trying to be funny?) I must be utterly insane. Maybe I didn't die after all, I just cracked up and lost my marbles, and this was the white room where they put the mentally ill people who were lost to the world. I could be wearing a straight-jacket and not even know it.

The sound of laughter reverberated throughout the room like a drum roll. A beautiful sound. I laughed too; at least I think I was laughing. I felt like laughing, crying, and hugging someone.

"All in good time, Anna, and I will remember your desire to hug me."

I had so many things I wanted to say back, but no voice of my own to speak aloud. Although I was pretty sure he could read my mind, or most of my thoughts, so maybe I could communicate after all. This was wonderful. I had so many questions to think, so many answers I needed to have, or I guess I just wanted to know. Nothing was truly necessary for survival, was it? Not likely, since I am irrevocably dead. I will be patient and wait for my form to take shape. I sure hope I am going to be pretty.

The laughter rang again, softer, and more distant, but no less wonderful.

Flooded with a renewed sense of optimism and happiness, I waited for the voice I knew I could count on, and for my body to re-grow. I noticed the table now held a writing pad, a pen, and small leather-bound book the size of a bible. Funny that I would be getting a lesson in religion at a time like this. I would read the whole book from cover to cover if it would yield some answers. (And if I had hands — darn it!) As it does with all lulls in events, my mind wandered. If I had been able to wish myself away, wish for light, wish to see my funeral (cringe), and wish to hear a guiding voice, then maybe I could wish for a few more things, like the ability to check in on a few people. I desperately wanted to know how Willie was taking the realization that I was actually gone. Had he accepted it yet? Worry flashed through me like multiple bee stings. I decided that I needed to know more than anything. I wished and wished to be taken or transported to Willie's house to take a peek. Whether it was good or bad, I wanted to see everything.

I concentrated with determination and 'willed' myself to concentrate on Willie's father's house. I envisioned the small white home, which could seriously use a few cans of paint. The five-foot-high fence surrounding the half-acre property was missing the few boards that their long-dead Rottweiler had repeatedly broken to escape. Inside the three-bedroom home, the clutter covered every table and filled every corner; dishes were piled on the counter and on the broken portable dishwasher. Willie's father had converted the front room into his sleeping, eating, and word-search puzzle area.

Since Willie's mother had left with his infant brother four years before, the house and its inhabitants had deteriorated. Willie had tried to contact his mother for two years after she left but to no avail. She obviously did not want herself or Willie's now four-year-old brother to be found. Willie had then given up and resigned himself to being mother and brother-less. Shortly after her abrupt departure was when the beatings intensified, owing to his father's drug problem, which had surely caused the demise of the marriage. The addiction increased, causing his temper to spiral out of control. Willie was his punching bag. I watched the changes take place in their family. Miranda showed pretend happiness to outsiders, while distancing herself from Willie and her husband. Then the baby came, and the facade dissolved. Then she was gone, and I was there to pick up the pieces that were my best friend.

Remembering the days spent hiding in the poster-filled room, strewn with dismantled stereos, VCRs, and DVD-players, made me happy. Sitting on Willie's bed, looking through his rock-band magazines, ignoring the Playboy under the mattress, we talked for hours, listened to music, and then escaped to my house when his father came home from work or woke up. We didn't talk much about what was happening in our lives, but we knew and we found comfort. Feeling the absent touch of Willie's hand on mine, I was suddenly there, on Willie's bed, staring at his tortured face. His faded black jeans were torn and filthy. The "Rage Against The Machine" t-shirt I had gotten him the year before was also soiled. His brown hair was thin, greasy, and clung to his forehead with sweat. He looked horrible. The shock I felt at actually being in the place I had willed myself to be again was quickly replaced by the total fear and desolation I felt at what I was seeing before me.

I reached out to touch him with my heart. He stared through me. His red eyes saw nothing, not even his walls. He almost looked comatose. This was more difficult than I had believed it would be. Willie was lost. Had I genuinely meant that much to him? Was this ravaged boy before me the product of grief over my death? Somehow, I knew it was. It was not conceit nor ego that defined my worst fear, it was the knowledge that I would be in his place if the roles were reversed. I couldn't just leave him like this, could I? Did I have a choice? Reaching out with my mind, seizing the connection we have had since we met, I reassured him that I was here, that he just had to look for me, dream of me, talk to me and believe in me. He just stared on through. To my surprise, on the floor, beside the messy bed, lay his journal, the leather-bound book he had kept his most intimate thoughts in. It was lying open to the last page on which he had written. I didn't want to invade his privacy by abusing my improbable position, but the need to know what the scribbles said was too strong. I had never read a single page in the treasured book before. I only knew it existed by accident. I had been cleaning up Willie's room one morning as he lay sleeping soundly after one of our common late nights. Picking up his magazines, I came across his journal, just as he woke up. Yelling my name, he grabbed it out of my hand before his face softened and he explained to me what it was. He had promised that I could read it someday, but only if and when our futures allowed for it. I had no idea what he meant when he said it, but I knew I would wait patiently for that day to come. I had read much of Willie's music and a few poems in the past, and hoped I would find more in his secretive diary.

Remembering the treasured closeness we once shared, I looked over at the displayed, half-filled page. The print was barely legible, but having a few years of practice I was able to make out what was written.

> How many pages have I written about you Adrianna? This whole book is practically dedicated to you and to think you almost read it. I wonder how you would have dealt with what I said, how I feel. But now I will never

know, will I? Your dead Adrianna and now I am alone. You were all I had in this world. You were my best friend. I had so many plans for both of us. You were going to be a wonderful nurse and I was going to do anything I could to be what you wanted me to be. I can't get the vision of your cold body out of my head. It is all I see. I have nothing but memories of you. It's always been you. I can't live without you. I am dead too.

> Shadows fill my head and darken my eyes
> I only have my memory
> The black reveals nothing but lies
> Of the man who could have been me.
> The candle flame is snuffed out
> My dreams fading then gone
> I find no voice to shout
> It was you all along.
> My purpose for life is now dead and so I die.

"Willie, oh my God, you're not dead, you can't die. I am so sorry. I love you!" I screamed.

At that, he looked up, his eyes focused, and a shudder went through him. His lips moved slowly, although I couldn't make out what he was murmuring. Sound escaped his lips in a rush, and he began to yell with fervour.

"Adrianna, I can almost feel you here, but you're not — you're dead, freaking dead! I can't do this without you. I need you. Why did you leave me? Why?" Tears streamed down his cheeks, his strained muscles rippling through his shirt and in his neck. I recoiled quickly. He was out of control. He was so lost. The pain it inflicted was worse than anything I had ever experienced. My death was killing him. As the entirety of his devastation hit me like a truck, I felt a mental tug on my mind. It became stronger with each passing second until Willie faded away and I found myself back in my fabricated white room. Nothing had changed but for the intensity of the light. The soft glow of candles replaced the fluorescent lighting. They were everywhere: on the table, on the floor, on shelves on the walls. They seemed to multiply before my eyes. I preferred the candle light, though it did nothing to ease the last image of my best friend and the brutal sensations that accompanied it.

"Anna, you can't do that again, you cannot go back there, there is nothing you can do. Please hear me, please listen to my words. You are meant to be here. Let the past go; it has no binds on you now." Even though I knew these words were meant to console me, they did not. I was relieved to hear the gentle, caressing tone and the sentiment behind it, but I couldn't accept them as being real. And yet this new reality (albeit perplexing and complicated) was now my life, my existence. I could not be there for Willie. It was I who put him in this state beyond repair. I had to believe there was hope for him, that he would find his way from the darkness and sorrow and continue with his preparations to leave his cage and become the person I knew he could be.

The desire to rip the pendant off my neck now sent a current through my mind and my soul. It seemed to weigh heavily on my neck and strangle my throat. I reached out with my mind which was now (holy crap) connected to a hand that I did not realize was there. My hand, my fingers, skin, and the small ring on my right pinkie that was there before I died, were all here. I could still see the fading remnants of the neon-yellow nail polish that Audrey had adorned my long, sculpted fingernails with, and the faint scar by my pointer finger. I could not believe my travelling eyes as they scanned my old body. I seemed to be intact. My tangled hair was still halfway down my back, my feet were still a large, unladylike size nine, and my long legs were still attached to my long torso, giving me my five feet and nine inches. I was still me.

Sadly, I was not wearing the clothes I had chosen on the day of my death. They had been replaced by a pale-mauve, long-sleeved cotton dress. My mother's choice, for sure. I would normally never be caught dead (no pun intended) in such a childish, delicate dress. However, here I was, looking like a little girl ready for her family pictures or her great-great-aunt's sixtieth wedding anniversary. I had a body now: my body, to be exact. There went my hope of being prettier than before. Could have been worse, I guess; I could have come back all rotted and gross. I quickly looked at my hands again to make sure this was not the case. Relieved that I was not decaying, I absently thought it would be amusing, and rather important, to have a mirror. I was sure I could feel foundation on my face and poorly applied makeup, including lipstick. I never wore cover-up, unless I had a massive blemish to hide, which was rare. There on the table, just inches away from me, was a small, white-framed mirror, suitable for a lady's dressing table. Wondering if I should accept the convenient apparition (it was totally cool), I peered at my reflection to confirm my suspicions. I had looked worse before. It wasn't as if I would be meeting any prospective husbands or going for a job interview. Being dead really lowered your standards of looking your best. Still, I had a hand, a sleeve, and spit, so I went to work removing some of the plastered-on make-up.

Elated that I had a physical shape again, no matter how ordinary it might be, I casually walked over to the white leather chair next to the table. Sitting down almost felt wrong. The thought of relaxing was a luxury that brought on guilt (honestly though, what else did I have to do?). Since I died, I had spent all my time staring, searching, teleporting, reminiscing, and apparently conjuring; it should be an opportune time to take a load off (even though I had just got that 'load' back).

"Your selflessness appears to have left you with sparse furnishings, Adrianna. I trust you have concluded that you may upgrade if you wish."

The voice was clearly different from the one before. This voice was higher pitched, not as rough, or as strict, and clearly the voice of a woman. It was not as soothing either. Turning to put a face to the interesting words, I saw a very tiny woman standing but three feet away. Startled that I had not heard the black high heels clicking on the stone floor or smelled the now-overwhelming scent of lavender, I took in every detail of my first visible person from 'the other side.' Wiry gray hair swept up into clips framed her lightly wrinkled but handsome, mature face. Eyes as green as a cloudy-day ocean scanned me from head to toe. I felt embarrassed as I X-rayed her in much the same way. She was dressed like a business woman ready to nail a giant company. The confidence radiating from her said she could easily achieve that and more. I liked her instantly; however, I felt unease at her drilling stare.

"Please excuse my intrusion, as well as my curiosity. You have been a much-talked-about personality for some time now. Your advancement into your form, mind, and circumstances is quite remarkable." She must have noticed my bewilderment and slight unease, as her face began to soften and she placed an aged hand on my arm.

"I apologize; you have not been told anything. You must feel exceptionally lost and afraid. I am your first visitor?" She asked warily.

My first chance to speak aloud to anyone since yelling goodbye to my mom (besides my ineffective attempt to yell at Willie) came out hoarse and inhuman.

"Yes, of course you are my first visitor; I didn't have a body to meet in until a few moments ago … or was it hours?" Her question was directed at me without the hope of an answer. She took it upon herself to answer for me.

"Here, Adrianna, we do not measure time in minutes, hours, or days, unless we are concerned directly with something time sensitive from the living world. I am not sure how long it has been since your body was laid to rest for you to be delivered here, though I am certainly glad it has. I, sadly, am not here to explain everything to you, although I can imagine you have many legitimate questions and concerns. I am here merely to meet you, to enquire about any previous visitors, and to tell you that you will have other visitors who will be much more explanatory than I." With this statement, she turned to leave through a door I had yet to locate. Honestly, I was slightly infuriated that I had gotten nothing but riddles from the strange woman who had just marched into my new room, and to top it all off she seemed to be a bit…well pretentious.

"Wait, please, can you tell me who will visit me and when?"

Knowing this was a futile question, I asked anyway. I did not want to be left alone again.

"I am sure many will wish to visit you. Many will have vital things to tell you, some of which might be difficult to hear. I am a 'seer'; I can tell you things that have been set in motion before they come to pass. But not everything is set in stone, Adrianna. Things can be changed if there is timely intervention. I have been unable to see much in your future — I am not sure why. I cannot answer any more of your questions. You will not be left alone long, I am sure of that."

She left after this. There was not much more I could say to make her stay. The unsettling feeling she left behind put me at a loss for words. I would enjoy having my body back, and wait contently for my next visitor and some answers.

* * * * *

"I do not see, Jonathan. As I told you before, I cannot seem to see any events that surround her anymore," the woman explained impatiently.

"Cripes, Valerie, you saw our world dissolve, and many have already left because of this. You know she is exceptional; you saw her as a child. How can she now set you to blindness?"

The question was meant to be rude, as Jonathan felt anger at the woman's conflicting stories. She was the 'seer,' the one who had foretold the unfolding disaster; she also knew that Anna was destined to be the one to save them, and now she did not know if it would happen. It was all so frustrating. Jonathan needed more to go on. He needed a vision to prove that breaking the rules was necessary, before the council deemed him a radical and forced him to move forward.

"You are testing my limited patience, dear boy. My visions are not exact, nor can they be manipulated or pushed as you have pushed our young friend. Gambling with her soul was not very intelligent of you. If she were not as resilient as she is, she could have been lost to us all. It is nothing short of a miracle that she did not go insane, being trapped in the dark, having to endure the pain of her very own funeral, and seeing her best friend fall apart, all the while being conscious without a form. The amulet is designed to guide one to one's destiny safely and is not to be given until the time is right. She is strong, son. That could prove to be a hindrance as well. Mind that you're aware of your own intentions here." The woman lectured with authority, standing over the top of Jonathan as he lay on his study couch. She had entered his sanctuary without greeting or permission. The elders always thought they could do as they pleased. She had started by giving him a synopsis of her visit with Adrianna. He felt perturbed that she had seen her first, even though he knew neither of them was ready to meet. This caused further irritation, which he took out on Valerie.

"Hey now, Jeez! Obviously, I didn't know she would be able to "will" herself to her own funeral, or anywhere for that matter. It must have been the Amulet that enhanced that need and power in her. I will worry about my intentions, Valerie; you need not remind me of the dangers here. I may be young to you, but I am not dense, nor did I die yesterday. I understand the limitations of your gift, and I hope that your visions become clearer. I am also acutely aware of what Adrianna has been through and how incredible a soul she must be to have succeeded though her peril," he protested shortly, though he wondered just how well he would have tolerated what she had just gone through.

"She need not have suffered, Jonathan. It may have been a slower process, but the anguish she has endured might have consequences. I must leave and have solitude."

"I will protect her, Valerie, and thank you for visiting her and attempting to see all you can." He thought it might be best to bring peace to this unsettling dispute, as he knew that Valerie had an exceptional gift and needed serenity to use it.

"One last warning, my boy, be particularly careful when you visit Adrianna. You have known of and seen her for many years now. She has been ignorant. Not to mention that your little 'visit' with her at the school ended rather confusingly for her, didn't it? Or were you privy to or concerned about her feelings after you abruptly left? Regardless, she is fragile and harbours many disturbing thoughts. Her special selfless soul has brought her here, but it might compel her back."

With this last revelation, she was gone, leaving only quiet and unease behind. He thought long after she left of her words. Having to recall the mess of a 'first impression' he had left Anna with that night was a sore spot for Jonathan. Sure, he had been told not to show himself to her many times before, but that did not lessen his desire to be physically close to her. The effect of touching her astonished him. He knew he had been unnecessarily rude, his words confusing and upsetting to her. There was no excuse for his immaturity, but for the fact that he could not seem to contain his energy when in her presence.

Knowing that Valerie was right did nothing to ease his troubled thoughts. Remembering many of the emotions and memories Anna had unknowingly showed him gave him a fear he did not expect. He hoped time would be enough to soften her pain, help her to forget this Willie, show her the glorious existence she would have, and bring her the friendships that she needed to forget her past.

Whether it was right or not, ready, or not, he was going to re-introduce himself to Anna. Making his way to her fabricated room, nerves tense as steel, he wondered how she would react to him. Would she recognize him right away? Would she be afraid? The longing to touch her soft velvet skin was distracting his thoughts. He envisioned her unusual, mystical brown eyes, knowing he could lose his soul and his objective if he weren't careful.

Jonathan began to make his way to her room when he felt a change in his surroundings. Knowing that Shadow was right behind him made his impatience flare. He just wanted to see Adrianna, not to stand here in the oblivion and endure another 'how-to-enlist-the-enticing-dead-girl-to-save-an-existence-that-she-isn't-even-aware-of-without-falling-in-love-with-her' lecture. However, he knew the meeting with the council must be over, and the old man might have some crucial information to share that might aid him in his quest. Turning slowly, he worked hard to remove the look of irritation from his face before he gazed into Shadow's eyes. He was still transparent and appeared to be upset. Jonathan softened instantly but still felt slightly defensive.

"Holy Saint Pete. Shadow, you do not look well. Is there distressing news from the council?" Jonathan asked, praying the old man's look of pessimism was not due to his pending visit to see Anna.

"I have just been to the council, yes, and some things have been determined and decided. I had hoped we would have more time to settle things before we spoke, until Valerie told us that she had seen you on your way to see the girl," the old man informed him. 'Damn, it was both: bad news, and I was busted,' Jonathan thought sardonically. Knowing he had to be patient and respectful with his elders kept many comments from flowing out of his restrained mouth. On top of that, he genuinely did like the old man. His being the initial, the very first, being to stay in the middle world gave him an unrivaled authority.

No one, not even the man himself, knew just how he got caught in-between the living world and the next. He had been born in Europe in the year 1149 and had been a man of considerable notoriety. Some of the other elders said he was one of the four knights who killed Archbishop Thomas Becket in the cathedral. There were also stories that Shadow had been a monk in that exact cathedral where Beckett was murdered, and that was why he wore the long, black monk's robe. When asked, Shadow simply declared that he had lived many lives ago and did not need to relive his life in the telling of it.

"The council and I have agreed that this situation we have has become exceedingly grave. Those of us who have been here the longest are fading, becoming transparent, as you see here." He indicated his changing form. He was still visible; however, so was everything else when you looked through him.

"We are also losing our ability to communicate effectively." In this, he was referring to the elders' ability to converse telepathically. Only a very few of the old ones could do this, and no one else seemed able to discover how it was done.

"I will not argue that your insistence on bringing Adrianna here, possibly before her time, and with such rapid methods, may prove to be more beneficial than I originally thought; however, with Valerie's inability to see the girl's future or use to us, we were unable, officially, to condone your persistence. That being said, we have decided to allow you back on the council to attend our meetings and to provide updates on her progress. The council still has lingering beliefs that you will have an issue with your patience and honesty, and possibly your 'feelings' when it comes to her, and that your judgement will be clouded." He paused for a moment to allow this all to sink in.

The fact that Jonathan was placed back on the council had everything to do with his connection with Anna and nothing to do with their actual desire to have him there. Despite their change of mind, he knew he would not attend a meeting if he could help it. His being the youngest, in life and death age, member of the council was like a burr in their britches. But being one of the few who could communicate with people in the living world while they were alive made him an automatic shoo-in. He was valuable to them, so they tolerated his abrasiveness. The old man put up his hand when Jonathan began to protest.

"We have concluded that our existence here is coming to a rapid end. Either we move forward and rid ourselves of our connection to the living world we all once knew, or we allow our world to dissolve around us and dissolve with it. Each one of us will have that decision to make, and we are unsure of how much time we have. I am sorry that this is happening, and I truly wish I could tell you why it is, but I cannot. There are those who suspect that there are darker forces that could be the cause, but there is no proof to substantiate this claim."

Jonathan did not wait for another pause in Shadow's grim words before he interrupted. He had to make the old man see that not all was lost. They had to trust him. He was important to this dimension as well, it had been foreseen.

"Whoa, let's rewind a bit here. I understand why you are so negative about all of this, considering you are the one who is fading away, but have some faith here. Valerie saw that I had gifts and that I would be necessary at one time, and, honestly, so far all I have done is tick everyone off, so my day must be near, right? And Valerie also saw, and so did Hunter, that Adrianna is supposed to be the one who saves us." He rambled with fervour. "Well, so far life has been pretty comfortable around here, and we sure haven't needed much saving, so I am wagering a bundle that she is meant to save you and me, and the rest of us middle-worlder's from having to dissolve in any way. Valerie is never wrong. Yes, I know things do change, and she has had trouble 'seeing' any more about Adrianna, but that just means it is now up to Adrianna herself and up to me to tell her what is going on. So, untie your shorts and let me work my charm, Shadow; give Adrianna and me a chance before you declare a state of emergency. I mean, you're not going to move forward right away are you?" Jonathan studied the old man's face for signs that he had gone too far or said something way off the mark, but the old man looked as if he were absorbing and reflecting on all that had been said. He hoped that he had not offended him, and that he would see that there was a probability that this could all go the way Jonathan was predicting.

Moments passed, seeming like a lifetime. His eagerness grew, but halted with the tears that crept into Shadow's eyes. He lamented being so flippant with his words as he realized the old man must be afraid to lose all he had known for almost a millennium.

"No, Jonathan, I had not planned on moving forward. If we are destined to have our existence dissolved, then that will be my fate. I never intended to leave, as Hunter and many of the others already have. I am certain of your conviction that Adrianna, as well as you, is meant to be 'special' in some way to all of us, but I do not know if this is something that can be stopped or changed. My prayers will go with you; may you be right. If I can be of assistance, I will be here for both of you." He spoke slowly, with a resignation that saddened Jonathan to the core and made him pray like no other prayer that he was capable of doing what he had just preached with so much assurance.

"Thank you, Shadow, I won't let you down. Please try to influence the council to remain, and I will report after I have given Adrianna a bit of time to assimilate all of this," he said with reverence. The old man nodded, smiled briefly, and vanished into the gloom once again.

Jonathan was shaken, his obstinate poise weakening with trepidation. How did he know that Adrianna could shield them from the unknown? He could not even tell her how. Was she to know? If Shadow did not have the answers, then maybe there were no answers to be had. Maybe this was their fate. Then what was Adrianna's fate? To be brought to this reality only to die again, an immortal's death? It couldn't be. They both had to have a purpose. He would not stop until he had exhausted all possibilities. He entered her room through the doorway she did not even know existed. Unbelievably, content as she was to tolerantly bide her time until directed otherwise, she had not looked for an outlet.

She was sitting on the white leather chair, her back to him when he entered the room. He could see her wild, black hair floating around her delicate but strong shoulders. The apprehension he suddenly felt almost made him turn around and leave before she even knew he was there. Knowing it was partly due to being unsettled already, he used that to give himself courage. He could sense calmness radiating from her. She must be thinking happy thoughts. 'Good job,' he thought as he slowly approached her. Trying to keep himself focused was becoming increasingly difficult the closer he came to her. Excitement filled him to the point of explosion.

She must have sensed his presence, as she turned in her chair, smiling at the incoming visitor. Still smiling, she squinted to see who was there. Beauty radiated from her face like from an angel's, and her chocolate-brown, intelligent eyes considered him without immediate recognition. Tentatively, he walked closer, interrupting her gaze, to lean against her table. Seeing the lit candles and the beautiful handwriting on the writing tablet made him smile. She was calm. She had accepted. She was everything he knew she would be. She was miraculous. Although he felt an edge to her acceptance, he ignored it.

"Hello, Adrianna. It has been a long time." His rough voice was a bit shaky and loud in the nearly empty stone room. Although it was warm, he found himself to be quivering faintly. Praying she wouldn't notice, he frowned slightly at his own reaction. He needed to stay confident, but her aura, even after all she had been through, even after her death, was breathtaking. "Do you remember me?" he asked hopefully, not wanting to be crestfallen if she did not recall their one-and-only meeting. A look of astonishment crept across her face as her pale hand rose to her open mouth. The silence was deafening. Immediately, he regretted coming in. He had no intention of upsetting her. Misery replaced his excitement as he prayed she would speak.

"You! Are you serious? Are you really here? Did I dream you up?" she asked, almost gasping for air. Watching her tremble stifled the laugh that was rising in his chest over her reaction, though a slight smile escaped, twitching the corner of his mouth.

"Umm... no, you're not dreaming, or having a nightmare. I am really here." Keeping his answers short would hopefully give her more room to speak and him less room to say something stupid. He quivered inside when she bit her full bottom lip and squinted again, this time in confusion. He noticed her right eye narrowed a bit more than her left, and her brow wrinkled lightly with concentration. Great! A few more adorable attributes to add to the list. 'How will I keep from kissing her?' he thought. Waiting for her to say something, he conjured an image of Shadow's worn face to fill his mind, bringing with it an uncalled-for irritation that he fought to suppress. He reminded himself that this was not her fault and that her obliviousness was essential until it was time to shatter her contentment.

"Oh my God, you're not here, you can't be, I can't be, I don't really see you, and this is all a figment of my deranged imagination. Crap! I must be in that nuthouse after all!" she exclaimed loudly, bringing her tiny hands to her eyes, viciously trying to rub away the image before her. When she started loudly humming an obscure song, Jonathan knew they were in big trouble.

5

Visits and Willie

When I open my eyes, he will be gone and I will finally wake up to find myself eating spaghetti with tomato soup for sauce with a plastic fork in a room where half the residents are singing "Somewhere Over the Rainbow." I removed my hands, and a dim light filtered through my still-closed eyelids. I knew I had to open them just to make sure I had been hallucinating, because surely I had been. My vision was slightly blurred at first but then cleared, not to reveal a large nurse standing over me with tranquilizers but that gorgeous boy with that same frown that rudely stole my heart at my high school. Ok, he is still here. So, does this mean he is dead too? Or am I not dead at all and I just hit my head hard when I fell on the bleachers, and he was just waiting for me to regain consciousness? I knew I had to say something and wipe off the idiotic expression I must have on my face.

"So ummm...wow...sorry about that, I'm just a little confused. Are you dead too? Or am I not dead, and this is all due to swelling in my brain?" It was the best I could come up with short of banging my head on his solid, muscular chest, or the floor.

A smile brightened his face, and a shaky spontaneous laugh erupted from his beautiful mouth, revealing perfectly sculpted lips, slightly crooked white teeth, and a small dimple in his right cheek. Oh my, if I am dead then he is an angel — a rude one, but still a heavenly creature.

"Whew, baby, I thought I'd lost you." He said with an enchanting smile.

"Adrianna, we have much to discuss, and I have much to explain, but you need to stay calm and relax. You are dead, at least to the living world, as am I. You have no swelling in your brain. We are in an extraordinary place that we like to refer to as the middle world. We are not living, but we still have a connection to the living world." He spoke kindly and slowly, probably hoping he would not bring about another session of frantic humming. He seemed to be treading very carefully, as if he had some earth-shattering news to tell me. Not sure why he would be so worried about a small freak fit when I should actually slap him for scaring the life out of me (yup, bad joke).

"Oh well, sure," I retorted sarcastically. "That all makes sense. Sure fits everything that has happened: I died on a sidewalk; I have no idea what I died from; I was trapped there, wished myself into my very own funeral to get away from the darkness, and now I am in this room. I finally have a body to move with, and I'm being visited by a woman that can see the future that I didn't know I had, and now you come along, the very guy I never thought I would meet again, who was awful rude to me ... what? Two years ago? I can be calm and relaxed ... sure." Although I felt embarrassed for running off at the mouth, the astonished look on his face made it humorous enough to keep me from blushing — well, at least not blushing as much as the first and last time we met. I couldn't help but take advantage of his stupefied silence, so I continued.

"So, you're telling me we are both dead, stuck in this middle world, still connected to earth, but what? Not allowed into heaven?" I had to ask. It was a curiosity of mine to see if I would be considered worthy to walk through those pearly gates, or if the fact that I had abandoned my friends and family earned me a spot elsewhere. Besides, I needed to say something that would prompt an answer from him, to remove the shock from his adorable face. When he saw the humour in my eyes, he softened and answered.

"You have certainly been through a lot, and the way it happened for you was unusual for sure. Most just die and do not regain a sense of consciousness until their body has been laid to rest, or cremated. However, you were conscious before your soul even left your body. I don't think I was responsible for that. The decisions you were able to make again are astounding. I am not sure I could properly explain that to you. But it is all irrelevant now, as you are here and we have finally met again. I do apologize for my rudeness when we … ummm … first actually met. It was the first time I had communicated with you and I was … well … a little shy?" He ended this statement with a question, hoping I would agree that he was just shy and leave it at that. Well I was not ready to leave it at that. He was boggling my brain with all this spiritual stuff; I wanted to be angry and keep it clear for a moment longer.

"That was shy? You didn't seem too shy to me. You flat out indicated that you didn't think I was going to make it on the bus safely and that you didn't want to talk to me. You said my name then, too." I remembered it all now like it was a few days ago, not a death and two years ago. "No, wait — you said 'Anna,' my nickname, which NO one uses anymore, and you left without even giving me your name." I didn't add the part about his saying we would meet again under not so easily fixable circumstances. I think I already knew what he meant about that now.

"By the way, would you finally tell me your name, since you obviously have known mine for quite some time?" Just how long he had known of me would be a question he would need to answer soon enough. Not to mention the fact that somehow he knew I was to die (way too young). Right now, I wanted to have the name of this miraculously baffling boy who seemed to pop up at the most interesting times.

He smiled again, and my heart jumped in my chest. Well, at least that is what it felt like, seeing that I wasn't sure if I even had a heart anymore. I also wondered how old he was; he certainly looks about my age, maybe just a bit older. It sure would be funny if he were 20 or something, remembering when my mom found out about Sebastian, the 22-year-old second-year mechanics student who took me to see the last Harry Potter instalment. His unusual voice broke my memory. "My name is Jonathan Michael Williams. It is extremely nice to meet you, Adrianna Marie. I was 18 when I died. Now may I please finish answering your questions?" I nodded for him to please continue.

"I was not supposed to talk to you, and I didn't expect to ... ah ... I didn't think I would have so much trouble conversing with you, so I figured it was best that I just leave. Then you looked so ... um ... if I may say, disappointed, and that bothered me, so I became angry at myself, which I guess took out on you." His eyes bore into mine with passion. I thought I saw a hint of irritation flash rapidly through his stare, which saddened me as quickly. Maybe I was being too harsh: he had said he was sorry, and he did come to see me, and he was the best-looking guy I had ever seen — well, right up there with Robert Pattinson and Paul Walker, that is. I waited to see if he would keep going, but he was silent, so I figured I had better do some backtracking, just in case he might decide not to tolerate me anymore.

"Yeah, well, ok, you're forgiven. So why were you at my school that night anyway, and why are you here now?"

"I went to see you. I had been sort of following you for most of that day, and I saw you ... ummm ... fall, so without really thinking about it I just allowed myself to be seen so I could help you. Now I am here to welcome you and, well, again just to see you." He was stammering like a cold motor trying to start. He seemed to be having a difficult time with his words, and I was a bit nervous that I had caused that with my rudeness. I wanted to reach out and touch him, to reassure him that I was not actually mad anymore, just confused and honestly quite scared. But I couldn't touch him. Shock was invading my body as his words sank in: 'I went to see you' and 'I had been sort of following you,' he had said. So, he wasn't there that night for the b-ball game, he was there to see me. Why would a guy like him, so obviously above my league, descend from the heavens just to see me? I knew there had been something angelic about him; I just didn't know how accurate I had been. There were so many questions I wanted answers to, but I didn't want to frighten him away. I felt comfortable now sitting so close to him, as if this were where I was supposed to be.

"You were following me? For how long? How did you even know about me or find me?" Starting with a few painless questions first might butter him up for some real slammers afterwards. He appeared to be contemplating his answer, and his face took on a surreal glow.

"Adrianna, I ask you to please believe that I will answer your questions as best I can, but I don't want you to become overwhelmed. We have time for me to explain everything to you and for you to discover all you need. So much has happened unusually rapidly for you. I am, ummm . . . worried." He paused, his delicate brows pinching with the spreading look of concern. He really was worried about me. Ha! I die, and this is what happens. Who would have ever thought?

"I have known about you for many of your living years. Since before your mom stopped calling you Anna, which, by the way, is a beautifully fitting name for you. I cannot explain to you just yet how I found out about you, but I can say that I am glad that I did. I was following you that day because, well, I had a strange feeling about you that day and I wanted to be there to help you. I had no idea it was you tripping over your lace, but I was glad to have helped you just the same. Especially considering the first bus, the one you missed because of me, got into an accident and a few people were seriously hurt."

He paused for a moment, and I was glad for the silence. He had saved me from a bus crash? Wow. This was all so utterly crazy.

"Anna, I have followed you many times, and I have learned so much about you even though I know there is so much more to learn. I have watched over you as much as I could, against the advice of some, and I have tried to keep you from harm. You are very distinctive, Adrianna, and soon you will find out just how special you are".

* * * * *

The speech, which he had practiced, was being delivered with more ease than he had expected. It was also more genuine than he had known it would be. He felt wrong at first, choosing his words so carefully to appeal to her heart, her undying kindness, and her soft spot for a well-played sappy romance. Now, as the words flew so easily, he wondered if he had left his script and was speaking from somewhere more intimate. It was difficult to stare into her eyes, even though he longed to; they had a way of boring into your very soul, undoing the complex bonds, and taking it into her own to see the world from her perspective. She had a mind like none other her age or any age. Compassion, optimism, empathy, sympathy, and passion were but a few of the emotions that ruled her heart and psyche.

Her soul was old, not in age but in wisdom, intuition, and the passion of her beliefs and her affections.

He knew many of the questions she had, so instead of allowing her the opportunity to ask them aloud, and potentially downright disrupting his train of thought with her captivating voice, he continued relentlessly.

"When you have had enough time to absorb all that I tell you, then I can safely involve you in a more intricate discussion about how you are such a significant person. Well, besides the obvious." He grinned, allowing the flattery to take hold. She smiled and blushed nearly as fully as she had when he had met her.

"The obvious?" she asked self-consciously. Her soft hands were fidgeting in her lap, nervousness showing in her every move and sigh. He knew he couldn't lay this on too thick for fear she would not believe him to be honest, and as dumbfounded as he was to acknowledge it, everything he was about to say was 100 percent how he sincerely viewed her.

"Anna, the obvious is how everyone you knew saw you, the way many here see you and the way I, never more than now, see you and believe you to be. You are gentle, loving, sweet, and selfless. You think of others before yourself, and you would assist anyone, at your own risk, who had even a hint of deserving your precious time. There are very few genuine people, living or dead, but you are one, and that is just a minuscule part of what makes you special."

The look of bewilderment that crossed her beautiful face just accentuated the facts he was trying to put across. Most would experience a puff in their sails from the flattering comments, knowing somewhere inside that they were true and should be told. But Adrianna just stared, baffled at his eloquent words. She did not see herself that way, nor would she ever; Adrianna was just Adrianna, the girl who had grown up mostly alone, the girl who gave herself to everyone she could, the girl who had more power than she would ever believe.

He wondered if he had said enough or too much, as she was still quiet. He wanted to spill it all, tell her every little thing that had happened to lead up to her death, what was happening now, and why they needed her, even though he was not even sure what they needed her to do. He just knew it was her they needed. Struggling to keep patient and reserved, he was startled to see that tears had formed in her timeless eyes.

Coming dangerously close to breaking a boundary he had placed on himself, he knelt next to her chair. Wanting to take her hand and remembering the two times he had touched her in the past brought on a feeling of overpowering weakness against her magnetism. He wanted to hold her and kiss away each salty drop that fell. He must get control. Stiffening his posture away from her, he looked up again to meet her lowered gaze.

"Anna, what is wrong? Did I offend you?" He was afraid, torn with her outward show of sadness.

"No, Jonathan, no — you did not offend me. Your words are precious to me, yet I am confused that you look at me that way. You barely know me. Hell, *I* barely know me. I don't want to be special, really; I don't want you to regard me that way. I am just a plain, ordinary girl, who came from a difficult life to die and end up here . . . with you. I just want to know what happens from this point on, what will become of my family and my friends, and what use am I to anyone now?" The tears flowed freely now, down her cheeks and onto her mauve dress. She turned her head away from him in time to miss the pained expression he wore to rival her own.

Jonathan had hoped to please her with his synopsis of some of her most admirable qualities; he was painfully shocked that they had had the opposite effect. He had made her sad. Somehow, she did not want to be special at all. Abandoning all caution, he slowly got up and sat on the edge of her chair. Reaching out with a shaking hand, he lightly guided her head to rest on his chest as he wrapped his strong arms around her shaking body. Immediately, he felt a sensation wash over him that he had never felt before. Internally, he shuddered uncontrollably as he took in her warm scent, which was like apples and cinnamon.

They sat like this long enough for her sobs to diminish and her breathing to become slow and steady. He felt her rub her eyes, heard her clear her throat, still coddled in his safe hold. She moved enough to gaze up at his serene face. Looking down into her red-rimmed eyes, he searched in their depths for an explanation of her inconceivable uniqueness. Her dark hair, soft and bouncy, framed her beautifully pale face. Her lips were parted enticingly. Without a thought of caution, he leaned down and placed his full lips upon hers with an urgency he had never felt. If his heart had actually still been capable of sustaining his life at that moment, he would have feared a heart attack — never before had he felt such a powerful pleasure or exhilaration from kissing someone. Their mouths sought each other like fire spreading across a field. His hand reached down to cup her warm cheek as hers slid through the thick hair at the back of his head. A moan escaped from deep in Jonathan's throat when Adrianna's other hand caressed the front of his hard chest. She reciprocated the moan with one of her own when he allowed his hand to entangle in her luscious black hair. Never had he desired someone so entirely.

Jonathan wondered briefly if she had ever been where this was taking them, decided that she probably hadn't, and told himself he would not let it get to that point now, when she was so vulnerable, or ever, if he was smart. The nagging feeling in the back of his immersed mind told him he must stop, that only hurt could come from this intimacy if she ever came to think that this had all been a tactic to charm her, or worse: that he was just using her. He could no more let go of her soft face and tantalizing lips than he could save the world all by himself. He could only pray that she would be the one to push him away and prevent them from falling further into this sweet, loving abyss.

6

Question, answers, and magical places

Wow, this is what it is like to be kissed by someone who can actually kiss. I felt quite dizzy and not so sure of my surroundings outside of his warm, muscular arms, hard, chiselled chest, delicately persuasive, strong hands and of course his easy, moist, delicious lips. I didn't want this to stop, but somewhere there was a voice that disagreed with me, telling me there would be time for this later. Always trusting my intuition, I reluctantly let my hand fall from his silky hair and pulled slowly away from his captivating mouth. I searched his eyes for any sign of irritation or anger, but to my surprise all I saw was a look of contentment and relief.

"Ummm . . . thanks for making me feel better." Wow, was I ever adept at saying the stupidest things to this guy. What was wrong with me? Oh yeah, I had just died, lost everything I had had, and been passionately kissed by the best kisser I had ever met. I straightened up in my chair a little to ease the pain in my back. Not only had I realized that I couldn't feel my heart beating, though it should have been pounding out of my chest by now, but could become dizzy and feel pain, and I also realized that I felt tired. I actually felt like I could sleep. Are we supposed to feel that? Oh, listen to the "we" crap. Been with this guy for a few hours and I'm already saying "we," as if we'd known each other for years.

The questions started to flood back in after the forced vacation they had just taken from my mind, but when I stared at his face, my questions halted again. The look he wore was one that scared me, one I was not overly sure I wanted a clarification of. There was sadness and the urgency of a different sort of need. There was also a coldness there that I had seen that night outside my school. He had withdrawn, and I wanted him back— I needed him back.

He stood quickly after noticing my inquisitive gaze, and leaned up against the table once more. He appeared to want to leave, although I was not sure why.

"Are you . . . okay?" I asked tentatively, fearing his answer more than I should have.

"Yes, Anna, I am okay. I am sorry for my apparent loss of control. I had no intention of taking advantage of you when you were obviously so vulnerable, which was because of me too, so I am sorry for that as well. Maybe I should go." His words carried a pain equal to that in his face, and I hoped to find a way to make him see he was wrong.

"Jonathan", -his name sounded so eloquent and respectable- "please, do not apologize. I lost myself a bit back there; it is not your fault. There is a lot to take in here, with you being what you are and saying those bewildering things about me, even though I can only disappoint you when you see I am none of those admirable ideals. But enough sad talk; I would like to think happy thoughts: they seem to make everything brighter. Honestly, that is a proven fact, and this glorious voice once told me to concentrate on the positive things, so let's do just that." I winked at him, letting him know that I knew it was his voice I had heard directing my memories and granting me advice when I most needed it.

I was happy to see his pouting lips flow into a genuine smile as his gaze met mine. After what we had shared, I certainly could not let him walk out of the room with sadness of any kind in his heart, especially if it had been a result of my emotional outburst.

"One thing I would like to know, though, is do you . . . well, we . . . I mean, do I need to sleep?" I stammered again, sounding like a complete moron, trying to avoid using the "we" word.

At least he was chuckling now, even though it was at my expense, and the sound was musical, although deeper and rougher than you would imagine coming from such an angelic face and figure.

"Holy Macaroni! I should have told you this already. Sleep is not a requirement, Anna. Our hearts do not beat, but we do breathe out of habit from our lives, though it is not essential either. Most of us don't care either way. Our forms will react and act much like they always have since we were born, but some of the elders here have taken it upon themselves to cease all activities from their human lives, though I am not sure why. We are, here, what we were on Earth, with the same characteristics, personalities, feelings, emotions, wants, and needs, except for the necessity of self-preservation, although, honestly, that's the hardest one to give up. Most of us still act as if we need to preserve our lives, when in all actuality we cannot be killed or die by accident or disease. That being said, our minds do become fatigued, and we feel as we did in our human lives: exhausted."

He paused for a moment, to allow me to take in what he had described. It did make sense to me. If we are so used to living and being a certain way and we have so many habits, we continue them on after we die without needing some of them. I smiled and nodded, to let him know that he hadn't totally lost me this time. He continued.

"Your physical form will not actually shut down, or slow down as it does in sleep, because your body is not a living thing. What happens is our minds wander to a "safe" zone or a pleasant memory, and we just concentrate on that and nothing else. I am not sure if I am explaining this correctly but that is the general idea. When you are stuck in that memory or thought, time goes by around you, but it seems like just a short time to you." Concentrating on his actual words and watching his lips form them as he spoke was mentally exhausting. (Such a pleasant exhaustion.) I knew I needed to go off to that safe zone and not think about anything complex for a while. I wished he would stay there with me; however, I was sure he must have many other more important things to do than to watch me daydream, and I would not be able to fade away with him beside me. (His presence could be quite exhilarating too.)

"I am going to leave you for a while, Adrianna; not that I want to, but I have a few matters to attend to. You rest, and I will come back for another question and answer period." Before I had even a chance to smile, or thank him, he leaned over and placed his lips on mine once again. This time, there was no urgency, just a gentle, mutual longing to connect on an intimate level. Breaking the kiss quickly, he turned towards a door that I now saw, quickly glanced back, and then was gone. On his face, as his eyes connected with mine in the minuscule departing moment, I thought I saw a look of regret.

Mulling over all I had learned since I found myself in this white room was impossible now that I had the "kiss" to think about. How could I call what had transpired between us just a kiss? It felt like more than a kiss. I had been kissed by different guys before, a few times. Well over five times and fewer than ten. And well, nothing could compare to how Jonathan's mouth and hands had felt. I decided that this would be my memory, my own thought I would concentrate on for my rest. I didn't want to dwell on the other ones that had been popping up in my head, like the look of regret on his face when he left, the look of coldness after we had shared such a marvellous kiss, and the strange feeling I had that he was keeping something from me, something crucial that I needed to know.

The nagging thoughts were pushed from my mind when I turned to find a folded note sitting alone on the corner of my chair. I sure hadn't seen him place it there. Unfolding the small, unlined piece of paper, I read the words aloud:

Beautiful Adrianna, how I have missed you so.

* * * * *

The next few weeks (as measured by living time) were filled with much conversation and adventure. Jonathan had never had such a wonderful time watching someone else discover new places and new ideas. Anna had asked a million questions and he had answered the ones he could, avoiding the ones he was not ready for and fearing the ones that were left to come. Taking her to his special park bench, where he had first seen her, was more enjoyable than he had imagined it would be. He was reluctant to share that space with anyone, and no one really knew how much time he spent there watching the children playing, the families walking hand in hand, and the wildlife peering out when all was quiet. She had been there before, of course, many times, though he was almost sure she had never felt the true beauty of this place until he showed her. The descriptions Jonathan gave of the ingrained images he carried were enough to moisten Anna's eyes with a film of happy tears.

On this day, there was a foot of moist, packable snow covering the path, spruce, and fir trees, as well as the park bench. He wiped off a spot for them to sit just as a quickly constructed snowball flew inches from Adrianna's nose. She laughed but jumped out of fright, not realizing the airborne weapon was intended for the teenage boy advancing out from behind the nearest tree, and that it could not touch her. Jonathan thought fleetingly about how fun it would be to join in and how easy it would be to do so. With overriding concentration, he could materialize in the living world to be seen, heard, and touched by its occupants. Adrianna could not do this yet (as far as he knew), although he would bet in time she would hone in on this superlative ability.

The adolescents engaged in a full-on snowball war, laughing, cussing, and getting soaked. Their shouts could still be heard even after they had retreated out of sight. Their departure left the park still, and Adrianna even stiller. Touching her chin lightly, he drew her gaze from the ground to meet his inquisitive eyes. She closed her eyes and asked him to be silent. Mildly frustrated at being shut out as she closed the windows to her intriguing soul, he sat back and waited for her to speak. She drew in a full breath through her nose and slowly let it go from her soft, full lips. Jonathan felt a startling stirring in his stomach and suppressed the strong desire to touch her.

Each second she kept still was an eternity to Jonathan. Worry slithered into his mind like snakes through the grass. Maybe she did not like it here in this park, or she might be wishing that she could be one of those teenagers outside on this unseasonably warm afternoon, or, worse yet, she could be sensing his impure thoughts and desires, which seemed to present themselves frequently when he was near her. Pushing these, and all other thoughts of that nature, out of his head took some control, but was accomplished just as she opened her eyes, once again gazing into his.

"Jonathan, am I an angel?" she asked quietly. The sensation that came over Jonathan at that moment was enough to almost knock him off the bench. He took in a fast breath and held it. This young woman never ceased to amaze him. Just when he thought he had her figured out, or at least had a decent understanding of the way her complex mind worked, she said or did something that rocked the very foundations of his presumed knowledge. He had no idea how to answer this question. He knew she was an angel more remarkable than any he had heard of. Dierdre knew it too, as proven by the miraculous picture she had drawn, but he was afraid to tell Adrianna the truth that they all knew to be true. She didn't handle being distinct very well. Despite his fear, he decided he would not outright lie to her again—he had deceived enough. His response was weak and unconvincing.

"Ummm, well . . . yes, I suppose you are an angel to many . . . including me." 'What is wrong with me?' he thought with embarrassment. He could not seem to focus on his plan of composure, and he knew she would want much more of an answer than his pathetic attempt.

"I'm glad you think so," was all she said in response to his stumbling words and thoughts.

Jonathan knew the subject of angels would come up again; he might even save her the shyness and broach the topic himself. If he could handle the conversation in an abstract way, he believed he could handle it a little better than this attempt. However, when he thought of the term angel, with all its history, it was her face that invaded the typical icon. Shadow and Michael would be the two elders who could explain angels to Adrianna; he had never taken much of an interest in the word or what it characterized, even though he himself was supposed to have descended from an angel and that his very name meant "gift of God."

As the dinner hour began and dusk fell in the living world, there were no more random pedestrians taking a trip through the snow-covered park. Jonathan was often reminded of a line from one of his favourite childhood poems: "Miles to go before I sleep" seemed to be the curse of all the preoccupied adults who took the park as a short cut to and from work. Occasionally, a homeless person would stop to rest awhile under the canopy of the trees in the surreal splendour of the park before they were rudely hustled out by the city officers. The three playgrounds with swings, slides, sandboxes, and climbing ropes were enough to draw many children every day. It was in the largest of the three that they now sat, with the setting sun casting natural shadows and bringing a brisk chill to the air. He wondered if she wanted to leave.

Adrianna had been quiet for what seemed to be a long time. Her beautiful but pallid face was calm and relaxed, her delicate yet gifted hands folded gently together on her lap. She didn't appear to be sad, just reflective. Jonathan fought back the impatience that swelled inside him. He wanted to know her thoughts. Now that she had completely accepted her position in death, he was unable to invade her mind as he could before. He had no ability to direct his thoughts towards her, although he believed the strong intuitiveness they each possessed had been the basis of that rare gateway between them. The connection had been weakened since. Probably for the best, albeit frustrating. She abruptly spoke, confirming his earlier conclusion regarding his aptitude to "know" her and to be amazed at her unique perceptions.

"I know that I have been here before, actually many times, with Steph. I came here with Audrey and one of my old boyfriends. But I never really saw this place; it was just a park, a place to have fun, sneak a smoke, although I only did that once, cut walking time, and to throw snowballs." She laughed at the earlier display of just that. "I now see why you come here when you're upset, scared or angry. I know why it is your favourite place and I hope you don't mind but I think it is mine too. Heck, this is probably where you first saw me. This very bench. Those very trees, those snow-covered flower beds and water fountain." She glanced at him, warily, hoping he would not be offended by her claim to his favourite place and her assumptions about his thoughts.

Jonathan laughed aloud in spite of himself, stunned as usual after one of her revelations of feeling.

"Holy turtles, how did you know that …Um . . . how did you? Oh, never mind that. Of course, I don't mind, Adrianna, I would be glad to share my favourite spot with you." How she knew the reasons he came here and that their first unknowing encounter had been in this very spot, he wouldn't even guess, but one thing was for sure: he had better start telling her soon about why she was brought here and what she was destined to do, or she would soon have it all figured out by herself and then be as mad as a swatted nest full of bees that he didn't tell her first. Jonathan placed his strong hand on top of hers, closed his eyes and imagined them back home.

After that visit to their favourite spot, they visited frequently: during the earliest of mornings to see the birds, squirrels, mice, insects, foxes, rabbits, and sometimes even coyotes; and during the day to see the tireless youngsters, fatigued parents with and without their pets, cell phones, iPods, and cameras. The evenings seemed to be her preferred time, when the visitors had dispersed, the noise had lessened and the air was cool. The one post-midnight visit they made upset her when the police ushered out a fragile homeless lady and arrested four drug dealers.

After a trip to any of the other places they visited, they always came back to the park. However, the trips to the living world took energy, and lots of it, so both Jonathan and Adrianna found themselves mentally exhausted after a trip and needed to rest awhile. This kept their visits shorter in duration and fewer in frequency than they would have liked.

In between the always-pleasant excursions to "their place," they met many other middle-worlder's together, all of whom welcomed Adrianna with a genuine show of positive emotions, and compliments on her beauty and angelic smile. She blushed, and stumbled over her thanks and words of kindness in return. It was pleasant to see Adrianna reaching the hearts of everyone she met, though it worried Jonathan. He wanted to shelter her from those he deemed too needy, to keep her protected and, honestly, all to himself. He tried to keep her as busy as he could, to absorb her time so that she was always with him and he could watch over her.

Frequently, their discussions focused on life in the middle world. Jonathan explained how when a person dies their soul is brought either to the middle world or through to a state of reality that even the long-time middle-worlder's did not understand. Most did not stay in the middle world, by choice or by some unknown divine intervention. Children never stayed. The elders believed that the ones who moved forward immediately were reincarnated.

He told her about Shadow and the other elders (omitting the council), about Valerie, Juniper, and many of the ones she had met and had yet to meet. He then explained how the middle-worlder's could come and go from one room or place to the next by thinking themselves there; It wasn't instantaneous but, measured in time, it was quite quick. Most had favourite spots to hang out, chat with others, or just be alone.

She had an inexhaustible list of questions, but left plenty of room for reflection and to enjoy their adventures and experiences. One conversation in particular was very intriguing to Jonathan. They had been lying on the small sandy beach on the shore of what was named 'velvet bay,' for the softness of the sand that covered the quarter-mile beach, which rimmed the perfectly blue ocean. Adrianna had made an honest attempt to create a sand crab while Jonathan studied her nearby. Her laughter danced on the waves and brightened the sun. He was marvelling at her beauty when she abruptly looked up and said plainly, "You don't understand me, do you?" He was so taken aback by her question and its accuracy that he stared open mouthed until she giggled at his stunned look.

Getting a hold of himself, but still looking quite foolish next to her intuitiveness, he asked, "Why do you think that?"

"You stare at me in question and disbelief when you ask me about the ways I have been there for my friends and family. You talk a lot about some of the things I have done that I find to be minuscule and basic kindness, as if I had averted a world war or prevented the apocalypse. You scold me for having put myself at risk for others, when that thought never crossed my mind," she stated strongly with curiosity.

"Ok, Adrianna, maybe you are right, and maybe I am worried over your genuine benevolence. The expression I must wear gives my ignorance away, but not to hurt or upset you. I guess I should just ask you right out: why, Adrianna, do you go out of your way to help everyone you meet, despite the obvious risk? Why do you speak kindly to those who don't deserve your gentle words? Why do you allow your heart to be crushed and sought after by those who are not nearly good enough to be in your presence?" He cringed at the intensity of his own words, more vehement than he had intended.

Memories of some of the things he had "seen" in Adrianna's life came flooding back all at once: images of her drunken mother leaving her to grow up fast and without guidance, of her mother's boyfriends, who were never outright cruel to her but damaging just the same. The worst for Jonathan was one of the parties she had attended, where she had a few drinks and the guy who had brought her had a few too many. The jerk tried without success to get Adrianna to take her shirt off, and then, when thwarted, became angry, grabbed her arms, and roughly removed her shirt. She slapped him and he punched her back.

Jonathan had wanted to murder the boy in cold blood and make sure he never entered any happy or pleasurable state of reality, but he had not yet learned to physically manipulate the environment in the living world without being seen. The boy had been interrupted by friends and she could get away. She did not tell anyone or bring any form of prosecution, believing it to be the alcohol that had made him act that way. She stopped seeing him and forgave his shortcomings.

There were other, less-disturbing memories of Adrianna making allowance for others who had hurt her feelings, of her putting her sister and friends above her own needs most of the time, and of the seemingly inexhaustible amount of time and energy she spent bandaging, consoling, and catering to her friend Willie.

She was right; he did not understand her, and she had given him the opening he needed to ask for revelation. She stopped smiling but did not look angry. Reflection on his words flashed across her face as she formulated an answer to soothe him.

"I don't see it the way you do, Jonathan. I don't go out of my way; I just see someone in need and determine if I can help. I want to help—of course I do. Everyone deserves help, a chance, and a second chance maybe. You might see a person who has done wrong, a bad deed or an illegal act, but I might see that same person's actions as a negative consequence of a bad life, a difficult childhood, or a moment of terrible misfortune. Everyone is born with a pure heart, and terrible things happen along the way to taint the purity and misguide them. Without anyone to redirect or show them hope, kindness, and even love, they may be lost forever. I cannot always help, and I am nowhere near quick enough to always know what to do, but if I have something in me to give, I want to give it. There is no one who is not worthy of my presence. I am an ordinary person who has made loads of stupid mistakes. I am not special; I just believe in goodness in people, in life, and now I guess in death." She laughed at the irony of her own words and went back to her sand creation.

Jonathan thought she had never looked more endearing as she did just then. That moment when she simplified the rarity and extraordinariness of an authentic, kind heart, when she alone made him re-evaluate every moral and value he based his every thought on.

Her hands covered in sand, a light wind playing with her wild, black hair all around her porcelain face, wearing jeans and a black tank, she could have been wearing a prom dress for how elegant she looked. She was dazzling. If God did exist and he had a vision of what an angel should be, then she was it. They visited places in the middle world that even Jonathan wasn't aware existed. One place made Adrianna cry. Her explanation was that it reminded her of the elf village in one of her favourite books and movie series. Obviously, she told him, the one who had created it had been to New Zealand, had an incredible memory for detail, and hadn't been here very long. The pale, carved stone walls of the high-in-the-mountains fairy-tale village looked soft enough to slide off. The nearby waterfall was loud and mesmerizing. The river-filled valley below fed horses of all colors and sizes. It was enchanting. She was very happy here, and Jonathan was delighted.

Adrianna then gave a three-hour synopsis of the movie versus the book, revealing colourful descriptions, intriguing plots, twists and events, and complex character explanations. Jonathan wished he had seen the movies; maybe he could obtain the books here in the library, although it didn't seem like the type of book that Shadow would stock up on. While she talked, they toured the valleys and hills that added definite emphasis to parts of her depiction. She ran to a river nearby and proudly exclaimed that it was where one of her favourite characters had fallen, presumably to his death. But he hadn't died; he was pretty beat up, though. Energy radiated from Adrianna in waves with force. It was obvious she enjoyed sharing her positive experiences as much as, if not more than he did.

Jonathan listened with acute attention to her every word, tone, and pause. She was quite the storyteller. A thought crept into his mind and flowed out of his mouth simultaneously before he had a chance to politely interrupt her or to determine the potential effect of the statement.

"You're Frodo," he stated simply. She stopped walking (well, she had been kind of prancing like a horse) and turned to stare at him. Curiosity radiated from her aura, but thankfully not the instantly feared anger of insult. (She certainly looked nothing like the hairy-footed young creature that she had described.)

"What do you mean, I'm Frodo?" she asked quietly.

"Well, he was obviously content in his little world and content to believe he was ordinary. Nothing special happened to him—kind of a mundane and relatively safe life—then, out of the blue, he reaches inside his soul to find immense courage stemming from compassion, strength stemming from determination, and an ability to succeed and conquer evil coming directly from the love of all things good." He felt himself blush and quiver inside at the words of wisdom pouring from him. Where had this come from? He could only guess it was from spending so much time with Adrianna. No one he had ever known before or knew now would believe he had just articulated such passionate words. He scarcely believed it himself.

He was unable to gauge her reaction by her facial expression or posture, and this frightened him. Slowly, a look he seldom saw (but absolutely adored) danced across her lovely face and left a spunky sparkle in her profound eyes.

"Whoa, Jonathan, that was deep." She stifled an encouraging laugh, grabbed his hand and continued prancing. Moments later, after remaining silent they both opened their mouths to speak. This brought on a good bout of laughter until she took the advantage.

"So, you're saying I remind you of a four-foot-high hobbit with furry feet, messy hair and giant eyes?" She laughed heartily. "Ok, honestly, I can see the messy hair thing but I shaved my feet last month."

"Ha! I'm glad to know about the foot-hair removal; I was thinking of offering you a foot rub after all the miles we have put in today. But no, you're surely not four feet high, and I love your hair." With that, he reached towards her face to lightly tuck her thick locks of raven silk behind her left ear.

"However, not sure you have ever been told this before, but you sort of do have elf ears." Jonathan faked a dodge, and began running up the hill with her musical laughter following his every step.

7

Pain and time gone by

I passed off the comment about my likeness to Frodo with a humour I only felt on a shallow level. It was nice to make him laugh and smile. He was beautiful when he smiled. All too often he wore a pained expression and I feared I had done or said something wrong. When I would catch him staring at me, which he did frequently, I felt a wide variety of emotions flooding me. At times his cobalt eyes would reveal a lust fit for the sight of a supermodel (hey, I'm no beauty queen, but the flattery is there), other times his gaze would rip through my insides and send chills of impending doom down my spine. Most often, though, there was just plain curiosity radiating from him like microwaves. I felt uncomfortable being regarded as a foreign object, or worse, an extra-terrestrial, and there was always an underlying admiration that I could not comprehend.

Still, above all other elicited feelings, the foreboding that kept invading my mind (besides the fact he wanted to kiss me lots) was that there was so much more than he was telling me, that there was a purpose for me here, and that his unfathomable comments about Frodo and me, seemed to hold more meaning and accuracy than I cared to acknowledge. Although uncertainty and worry plagued my thoughts, I was still content to wait patiently for the moment when he would divulge the answers to my increasing questions, to be cognizant of the difficulties he must be having in the revealing, and to spend as much time enjoying his company as I could. (The last one impossible not to do—he was one fun guy.)

Deciding that I needed a rest, I asked Jonathan to take us back to my room, where I now spent very little time. I needed to not see, not hear, not laugh, and not worry for a short time. There was so much to accept, take in, and understand. The beauty of the places we visited would need to be remembered over and over to possibly comprehend its entirety.

I felt a strange sadness sweep over me when Jonathan told me he wasn't going to stay in my room; he had things to do. I knew I would not rest with him there anyway, and I shouldn't absorb all his time, so I conceded by giving him a hug, thanking him for the wonderful adventures, and apologizing for the long, drawn-out tale of Orcs, warriors, and a skinny goblin. He left after hushing my apologies and gently touching his warm lips to my cheek. The feeling of his touch lingered long after he left, as I recapped some of the adventures we had had in such a measurably short time. Once again, I was completely surprised to see another note from Jonathan to add to my growing collection. They were always short but lovely. The emotion brought on by just a few words made me feel positively giddy. This one made me laugh aloud, despite the confusion it created:

To my Frodo, rest well and know you are never out of my heart. You will rid yourself of that ring soon.

Love, your Sam

Despite the majestic beauty of the green hills, waterfalls, and empty city, my favourite place was still the man-made, modern park where he had first taken me. I felt a subtle but undeniable sentimental feeling that did not come from me. It was easy to see that Jonathan regarded that place with intensity and gratification. I suspected again that he had first seen me there with Steph. He had not confirmed this, and his reluctance had heightened my suspicions. Why it was such a memorable moment to have watched two silly girls playing on a nearby swing set, I had no idea. (I had been a gangly pre-teen with absolutely no boobs.)

However, the facts were these: Jonathan was obsessed with me, but (my ego crushed like a grape) I was sorrowfully confident it was not out of love; he needed something from me and was ready to say almost anything to get it; he was gorgeous, charming, intuitive, and extremely entertaining to be with (a much-desired mixture of Edward and Jacob, my favourite vampire and wolf), and I didn't want to be away from him; and my feelings were increasing despite the red flags springing up at each twist in the road. These were the particulars of my situation. Boy oh boy, this couldn't end well. I am dead; he is even deader, and we are stuck in a reality that doesn't change or evolve, seconds but light eons away from our loved ones in the living world, and he has a secret big enough to lock up, stamp 'top secret,' and bury in the sea bed.

With so many thoughts partying in my head, it was almost impossible to center on just one for the rest of my functioning brain to shut down. It wasn't going to happen. Nope, never, out of the question. I couldn't lock down. I had no trouble the last few times, as I had been overly bushed, but this time there was just too much. So instead of continuing to replay the recent past with Jonathan in my new existence, I allowed my mind to wander further . . . back.

I thought of Audrey, wondering if she and Nicole were missing me, still hanging out, and getting ready for Christmas. I bet that Audrey had at least reached first base with Darwin, the gas station jockey in the city center; he was cute, for sure (but nowhere near hot like Jonathan, or Theo James—just totally sexy). Nicole, the forever student, had no one lined up when I died. She was too shy to flirt and too sensible to go anywhere devious to meet any bad boys. I wondered what they would think of Jonathan. (Oh Yeah, he was edible, they'd love him.) Willie would like him too, though he wouldn't admit it, but they didn't have much in common except for a desire to protect me.

Willie. I hadn't thought of him much since the visit to his room on that painful day. How could I be such a jerk? He was my best friend, and I had just gone off on my new life (well, death) and forgotten him after the devastation I had left him in. Jonathan was so wrong about me; no genuinely good person would do that. Even Frodo under the influence of the ring didn't forget about Sam. (Even if he did go off and leave him—cringe.) Maybe I could check in on him again. He must be coping better, now that he has had some time to adapt to the fact I am long dead, buried, and gone.

I concentrated very hard on his face; his dark hair, which always needed a cut and style; his emerald eyes, bearing a hint of sarcasm and spirit; his tall, lithe body covered in scars and the one dragon tattoo he had gotten on a whim last year. I had wanted to get a tribal tattoo on my lower back (yes, I know: a tramp stamp) until I watched the needle stabbing a million times a second all over the right side of his chest. Well, I guess I didn't have to worry about that now, did I? I could plainly see him perched on the side of his bed, fiddling with his speakers, new DSI, or the many other electronic things he had scattered everywhere.

Somehow, it didn't feel right. The vision of him was almost right, but not his room. He was not at home. How I knew this for sure I didn't know, but I knew. A feeling of dread came over me as I reached with my mind to find him. He could be at school, or out with his other friend Jake. Why did I feel so apprehensive?

Suddenly I found myself in a dim alley next to an open dumpster. It was filled with what looked like wrapping paper, boxes, and plastic. A little early to be chucking that stuff out, wasn't it? A sign on the side of the graffiti-filled brick wall told me I was in between a 7-11 and a grimy hair studio in one of the grossest areas in the northeast end of the city. I only knew this place because I had been here once before, when my brother was picking up his girlfriend. I looked to my right and was confused to see a Christmas tree leaning up against a fire escape. How odd.

I couldn't fathom why he would be here, and I suddenly worried I had "thought" myself to the wrong place until I heard his voice, only it wasn't his voice. It was Willie, but his tone was low, rough, and angry. Turning in the direction it came from, I saw the back of a black leather jacket and filthy ripped jeans. A very tall, gangly man was standing slack by the wall facing my best friend. If it was possible, he looked even worse than the last time I had seen him. A black circle darkened his right eye and his lip was split with dried blood. This bruise was not given by his father but by someone left handed. Willie had been in a fight. He wasn't wearing a coat despite the bitter cold from the cloudy, snow-filled sky. The black, plain sweater looked like it was one of his father's, which shocked me. The hood was up but not enough to conceal his immobile, tortured face.

What stunned me, enough to go weak in the knees and lean up against the wall, was what was in his hands. In his left hand, he held a half bottle of what appeared to be whiskey; he didn't even have it concealed in a paper bag. In his right, he held a bag of what I was shockingly positive was marijuana. Willie didn't drink, and Willie was against drugs. Oh sure, in the past we had had a few drinks together and with other friends. But never, ever, had we tried or even wanted to try drugs. With my mom's history with alcohol and his father's progression from weed to cocaine, we had made a solemn vow to never go down that road. We had both seen, and physically and mentally felt the damage that could be done to the family of someone with such a lifestyle. No good could come from even trying something that was obviously illegal for a reason. Why was my best friend holding a bag of the very stuff that had destroyed his childhood and nearly killed him through his father's dependency?

He had obviously just bought this from the skanky character in front of him. How could he have gone from the stoic, strong, loving man I had known to this chemically dependent, lost boy who held two of the most potentially dangerous substances in the very hands that had so frequently caressed my face and held me safe such a short time ago? It had only been a few weeks, hadn't it? With a jolt, I realized that I could have taken for granted the time that had passed. The Christmas garbage, the used Christmas tree: proof that the holiday I loved so much had come and gone. Had I really been dead for over two months in living time?

On my visits to the mortal world, I had not stopped to take inventory or to re-acquaint myself with time that was measured in days and weeks. It was disgusting how easily I had conformed to my new reality and forgotten those among the living who survived each day within those boundaries and traditions that I had thrown away. I was responsible for this. I had let my best friend down. I should have been there each day watching over him. Jonathan was able (with much control and strength) to manipulate his presence in the living world to be seen and touched. I could have learned this by now. I should have asked him to teach me; then I could have prevented the decline of him who used to be the most important person in the whole world to me.

I nearly jumped out of my skin (crap, I hoped that wasn't really possible) when I heard his voice again, so close and clear.

"Sixty now, forty next week. I want to meet somewhere else, too. This place is shit. I got your number." The voice was the same, but not. It was cold and devoid of expression. This was not my Willie.

"Yeah, Will, give me a call when you want some more. Its good shit, man, and I got something else coming in that will give you a real kick in the gonads. You sure you don't want something with a bit more slam?" The unnamed man spoke in a high-pitched, whiney drawl. Obviously, a complete dope head.

"Man, I told you what I wanted, and I got what I wanted. If I want more, or anything different, I will tell you. Now piss off, will ya? I want out of here," he growled in his new, not-Willie voice.

"Cool, man, cool. Catch ya on the flip side." The man walked past me (completely oblivious to my presence, of course) without looking back, probably to take some other good person's money, or worse, off to corrupt some other young kid. After the sleezeball left, Willie stood there for what seemed to be a very long time. I guessed he hadn't really wanted to take off but wanted to get rid of the dealer. Staring off into nothing, fists clenched, he shuffled his feet but didn't walk away.

I wanted to reach out to him but, honestly, fear took a hold of me. Even if I could summon the power needed to speak aloud for him to hear, or to touch his rigid arms, I wasn't so sure I wanted to. If anyone had told me that I would eventually become afraid of my best friend, I would have told them to jump off a cliff (or maybe to just shut up). The reality was that I was afraid. There was no guarantee he would even want to know I was there. It might drive him further into the dangerous state he was already in.

He took a long drink from his tightly gripped bottle, his hard body swaying slightly in an unfelt wind. He's drunk. I guess I should have assumed that, considering the bottle was half empty a few moments ago, and now only a quarter full. Knowing I had to let go of my fear and at least try to try to communicate with him (this was all my fault) did not make the concentrating any easier.

I had no idea how to do this (Omigod, I am learning how to haunt my best bud!), so I stared at his impassive face intently and tried to speak aloud.

"Willie, It's me, Adrianna." Wow, did I sound like a moron, and an ineffective one at that, because he didn't move. If at first you don't succeed, try, try again (or lower your standards). Who the hell said that, anyway? Obviously not someone dead trying to communicate with the living.

Focusing harder, Jonathan's face flickering in and out of my mind, I held my useless breath and bore my stare into his vacant eyes.

"Willie, it's me, your Adrianna." Louder might be better. Nothing, not working at all. Man, I couldn't get anything right.

"DAMMIT WILLIE, CAN YOU HEAR ME?"

He flinched as if he had been bitten by a snake or stuck by lightening; his whole body jerked, his unseeing eyes darted around. Clutching his head, he let out a low moan and trembled. Taking advantage of my success, I spoke again, a bit softer this time.

"Willie, it's me, Adrianna. I'm right here, even though you can't see me. I'm working on it." I was naively hopeful that this would calm him long enough for me to continue to work on it.

The look of horror that crossed his face broke my heart as he gripped both sides of his skull with his now-empty hands. The bottle did not break when it hit the ground, although it sure did when he kicked it hard into the steel garbage container.

I jumped again when his words reverberated through the cool crisp air. "Get out of my head. I DON'T HEAR YOU—GET OUT OF MY DAMN HEAD!"

Cringing and regretting my poor attempt at consoling him, I slunk back as if I were afraid to be seen. Willie turned quickly and ran past me down the alley and into the dark bitter night. I slowly sat down in the snow-covered alley and cried. Devastation took me over until I could not feel anything around me. I was numb everywhere except for the pain that suddenly felt huge inside my empty chest. There was no heartbeat, only spasm after spasm of wretched pain. I knew then that if I could learn some after-world parlour tricks like Jonathan, if I had one purpose in this immortal world, it was to find a way to help Willie before it was too late. Before he ended up dead like me.

Dead like me? Willie could come to the middle world if he were dead. Whoa, that thought came out of nowhere. Willie might end up here with me and, of course, Jonathan. Oh, that would be awkward. But I would have my best friend back, and I was sure if I explained the situation, Jonathan would help and welcome Willie as well. I would definitely have to run this by him, though, when and only when the time was right; of course he would understand, and he would teach me everything he knows. But that was insane. I couldn't actually be contemplating killing my best friend to save him.

This was all too much. Everything that had happened since I died was too much. The longing I had to help Willie, the excitement mixed with fearful anticipation each time Jonathan touched me, and the crazy things that kept happening overwhelmed me to the point where I wished I could have just died and ceased to exist completely. I had felt sorrow, grief, pain, and misery in my life, but nothing compared to what I felt now. I could not be wishing that my best friend would die and that my new friend (I almost said boyfriend, oh my) would help me. No, this was all wrong. I just wanted to save him, hold him, hear his beating heart, and hear my own beating heart.

I had an eerie feeling, creeping up the back of my neck, that I could now thankfully feel again, and telling me that there was more pain to come before I found a way to fix the damage I had done to Willie. My intuition also told me that it had something to do with the secret that Jonathan was keeping from me. Well, now, we are both even: both with something to gain from one another (I would learn everything I could, whether he wanted me to or not) and an undisclosed objective in the process.

Just as I stood up from the chilly, wet ground, I felt a hand on my shoulder. Spinning around faster than I thought I was capable of brought me face to face with Jonathan. Even in the dim light from the cloudy, moonlit night sky, his face was extraordinary. So, smooth, pallid, and frankly irresistible (ok, he has damn fine kissable lips, and smouldering bedroom eyes). His eyes were not full of lust, however, but they seemed to be smouldering. He was angry, and there was something else . . . fear. Why was he afraid of me? No, that wasn't it. He was afraid for me. Crap, how long had he been there before he made me aware? Had he seen Willie buying the dope, my pathetic communication tactic, and my embarrassing breakdown?

I stared right back into his knowing eyes and realized he had. "Adrianna, we need to talk."

In a split second, we were back in my lonely room, which I had left not so long ago. Although he was quiet, the tension was pouring off him like a waterfall.

"What were you trying to do back there, Adrianna? No, don't answer that. I know what you were trying to do, but why? I don't understand. You have no idea how to materialize, no idea how dangerous it is for both you and that idiot you're driving insane."

I was gobsmacked, and that didn't happen too often. I could almost feel the sting of the verbal strike he had dished out to me. I wasn't trying to hurt Willie. I could not grasp his anger (apparently, even dead guys have way too much testosterone). However, my temper was rising to meet his.

"Excuse me, Mr. I-can-do-anything-and-you-can't. Willie is not an idiot; he is just . . . ummm . . . not himself. I am not trying to make him go insane. I have to help him. I can learn to materialize, I know I can. He heard me. Just like that day in his room, he heard my voice. If I just concentrate a little harder, then I know he will see me too. You could even show me how." I gritted my teeth at my last words. Right at this point I wanted to slap him more than I wanted him to aid me in any way, especially after the mean things he had said. Not to mention that I wasn't quite ready to reveal what I wanted from him before I had a way to push him into telling me what he was hiding. Everything had been going so well; why did he have to act like a jerk now, when I needed him the most? My anger was not subsiding.

"How is it dangerous, Jonathan? You did it with me, so why can't I do it with Willie? Oh, that's it: it's Willie, isn't it? You're jealous!" (Whoa, I really jumped out on that limb.) I wanted to take back those last six words. I had such a big mouth. I had no right to make a rude assumption like that. He wasn't my boyfriend, we had no commitments between us, and, well, I had no idea how I felt about him or how he felt about me (I'm sure he thinks I am twit now). Just because we shared an electrifyingly passionate kiss, where I seriously wanted to feel his lips more and more (and tear his clothes off), and he had kissed me sensuously on each short parting from each other, and I could not stop myself from loving every moment we had had together. Oh, crap. I would be right freaking jealous if I were in his shoes.

He stared at me quietly, an incredulous look on his face, lips set hard and straight. I cringed at what was coming when I saw his mouth open to speak. "I don't think my being jealous or not is the issue here." His anger was gone, resignation taking over. This I hadn't expected, and it caused my guilt to transform into tears in my eyes.

"Your friend doesn't seem to be doing an adequate job of keeping his marbles in line, Adrianna. If you insist on trying to communicate with him, or worse, you succeed in materializing in front of him, you will inadvertently push him over the edge to insanity. Then, if he commits suicide because of it, he will end up stuck in a room with many others who he will not see or be able to talk to. He will literally be bound by his depression for eternity. There is no hope for suicides, and their lives are surely better than their deaths." The painful shock I felt at his words, that I could possibly cause that to happen to Willie by trying to help him, was more than enough to unleash the flow from my straining tear ducts and keep me silent. He continued in his low, emotionless voice.

"I personally do not care what happens to your friend, but I do care what happens to you. I will not teach you to do something that is going to hurt you in the end. You must force yourself to move on, Adrianna. Let Willie go and concentrate on what is here in your new world. If you can't do it for me or you, then do it for him." Bitterness crept into his voice near the end, and the strained look was back. He was jealous, but I had no right to force that statement from him. He did care about me after all. He said he did, and his worry for me was real, I could see it.

Not having a single thought in my mind about what to say, I reached out and took his hand in mine. He reluctantly stepped forward to reduce the gap. His close presence sent a shock through my hand and into my body. Suddenly, I wanted to be closer. Much closer. Half jumping, half falling, I landed in his ready arms and held him with all my might. Tears flowed steadily now all over his black button-up shirt and warm, soft neck. I did not know how long we embraced before I felt his hot breath on my ear and neck. I turned my head to catch the expression on his face and his lips were instantly on mine.

The kiss was fast and furious, like before. There was a hardness behind his lips that was new and frightening. I had hurt him, belittled his feelings, and made him worry. I realized that our feelings for each other were growing, and he could feel it too. But what if I was wrong? What if he just worried that I would mess up his plans for me? I couldn't think. The heat was searing through me, and I was melting in his kiss and grip. I didn't want this moment to end; I didn't want to think about Willie, or Jonathan's secret. I didn't care what it was as long as he would kiss me, hold on to me, and protect me from myself.

Just as I slid my hand up his hard, muscled back, he released me. He stepped away quickly, a look of sadness stealing the passion.

"Please think about what I said," and then he was gone. For the second time that day, I sat and cried straight from the soul.

* * * * *

The argument from earlier was not mentioned or even hinted at when Jonathan came to get Adrianna for another adventure. The tension was thick enough to cut with a knife for the first while, but when she saw the waterfall that he brought her to see, she quickly loosened up. Her smile and laugh caught his heart in an instant when he carried her over the river to the high embankment. They climbed for what seemed like hours until they reached the cliff. Laughing at her hesitancy to approach the edge and add image to the magnificent roar from the frothy water, he felt himself lighten too. The picnic he presented was evidently just a humorous, touch but the strawberry-rhubarb pie, butter-lathered buns, and cream soda brought an even bigger smile to her beautiful face. She was delighted to find out that food still tasted great.

Pressing aside feelings of guilt over their last conversation, he was determined not to bring up any sensitive topics for either of them. After he left her the last time, the kid had plagued his mind almost as uncontrollably as the fear of what would happen if she kept up this pursuit to help the moron boy, who was obviously not nearly good enough to have had her as such a close friend. He was a drunk, an addict, and he had a seriously bad future.

Jonathan had sensed a very dark energy coming from Willie when he had found Adrianna watching him. He could not understand the feeling that had come over him, as he normally had no premonition abilities regarding random people in the living world—without the amulet, that is. Somehow, he had felt that there would be much more from this Willie character, and none of it good, although he had no idea what it was. Maybe he would speak to Valerie about it. He had to find a way to remove Willie from Adrianna's head. It was grating on his nerves. He didn't believe she loved him—well, any more than as a friend—but her feelings for this disturbed boy bothered him. Jealousy ate at his insides, driving him to detest one whom she cared for so much. In time, she would forget him; he would make sure of it. He had to. Maybe Valerie would see a way . . . .

They ate some of the well-prepared snack and lay on a blanket a few feet from the cliff's edge. Conversation was light and confined to the surroundings and the middle world. Her questions were much easier than before, she being still a little resigned. She asked whom he had first met when he had arrived, so he told her the story of Dierdre. He hadn't introduced them yet. However, he knew they would meet. It was providence.

Jonathan had been angry and stumbling around to try to find the 'seer' woman whose name, at the time, escaped him. He had heard that she could tell him things that no one else could, and he had been understandably desperate for answers. Out of the corner of his eye, sitting on the floor, was an angelically beautiful little girl about the age of five. He asked the first person he saw who she was. Her name was as mystical and rare as she was, and one he had never heard before: Dierdre. At least, that was the name on the inside of the sweater she was wearing when she had arrived.

She was smiling, holding onto a little stuffed puppy wearing a purple ribbon around its neck. Her silky, dark-brown hair framed her round face and gave her a pixyish look. When she stared at you with her startling, green eyes, you felt as if she had seen your soul and knew every thought, desire, and dream you'd ever had.

Figuring he would be different from everyone else and attempt to make friends with the little-one, he approached her.

"Hey Dierdre, how's things with you? That's a pretty cute puppy you have there. But not as cute as you. Have you seen Valerie here today?" He asked questions but did not expect a verbal answer, for he had been warned he wouldn't get one. Dierdre had not spoken a word since she had arrived. No one knew where she had come from, what had happened to her, or why she was bound here with the rest of them. She sat and watched the comings and goings around her, fascinated and always smiling.

Their next meeting was even more thought provoking than the first. Jonathan was again looking for Valerie, loitering around the common room, hoping she would flit in as she usually did. Dierdre was at a table, drawing pictures. When he had finished playing twenty questions with some of the long-timers, he went to sit in a chair in the corner by himself to think. Dierdre came over, sat down on his lap, and hugged him.

He felt a warm energy pass between them as they sat holding each other. Jonathan felt a brotherly love towards the little girl in that very instant. The feeling made him miss his own sister, and he wondered what would become of her now that he was gone. Dierdre left his lap and went to her table. As she was coming back, he noticed she had a single piece of paper in her delicate hands. She then showed him a remarkably meticulous drawing of a boy and girl, holding hands, and walking into the ocean under the shadowing moon. The boy looked like him, and the girl looked like no one he had known until recently. The beautiful figure had black, full hair and a stunning mauve dress. Adrianna.

Jonathan always felt a pang in his heart whenever he saw Dierdre. She was a rarity. It was a mystery to him, as well as to everyone else who had met her, why she had not moved forward. After seeing her pictures and spending some of the day with her, it saddened him to realize that she would never socialize with her peers. No one knew why, but children moved on immediately when they died, which supported the theory of many of the elders that there was a better, more peaceful existence to be had when one moved forward. However, where Dierdre fit into this theory he did not know. There had to be a reason for her to be here. He prayed for her that in time she would know what it was. Dierdre had become a timeless fixture in the common room long before he arrived, and quickly became an eternal figure in his heart.

Although he had left out the part about the perfectly depicted drawing she had produced, he felt he had done a decent job telling Anna his account of Dierdre and the ambiguity surrounding her. She seemed very interested in the sweet, mysterious little girl, as he expected,   and asked to meet her. It dawned on him that they had not come across her in the past two months, but then they hadn't spent much time in the common room either. They agreed to set that up soon. He put Dierdre out of his mind for the time being and concentrated on Adrianna.

Jonathan wanted to touch her again. Not as he had the few times when his instincts had taken over and he had lost control, but with the sensual embraces and gentle kisses they had so often shared. He wanted to hold her hand as he had in the park, the elfin look-a-like village, and the million other places they had toured, sliding his wanting lips across her cheek and playing with her hair. Fear immobilized his hands at his side.

After some time had passed, Adrianna sat up and started picking a few stray leaves off her silky, black tank and blue jeans, stretching her back, her hair falling softly to the ground. He knew he could not allow this day to end without the chance to caress her and attempt to erase from her memory the harshness of their last encounter.

Worrying that she was weary of their new spot, perhaps from her fear of falling into the raging river, he remembered one other place he had wanted to take her as soon as possible, thinking it was the place they needed to inspire some softness in each of them. And he had a surprise for her, and it would take one stop to make it happen. Positive she would be thrilled, the excitement bubbled inside of him like a pot of boiling water. Jumping up, startling her, he said, "I have just the place. It is close by, physically; we can actually walk there, if you feel up to it, and I just have to make one quick stop at the common room."

"Umm, yea, that would be great, walking . . . A very human thing to do." She laughed and patted off her blue jeans. He couldn't imagine her not looking gorgeous in anything she chose to wear, like the baggy blue jeans and pale yellow t-shirt with "Save the Environment; use paper bags" that she had worn the previous day.

Enticed by the energy in his posture and voice, she smiled and got to her feet. They walked quietly, taking in the almost-cloudless sky, the light, floral-scented breeze, and the eagles flying overhead, ready to swoop down, talons sharp, to their lunch. It was all so wonderful: like the living world, but without the hint of pollution in the air, the hidden dangers, and without the restraints of time. It was safe and relaxing. It was easy to see that Adrianna felt the same way. Her face was tranquil, her body fluid in casual movements.

They reached the common room. Anna was a bit shocked when Jonathan asked her to wait outside for just a moment. Tapping her foot with irritation, she pursed her lips and nodded once. He almost ran into the room in search of his special surprise. Finding Dierdre at the table that she drew her pictures on intrigued him at once. Maybe she would have another one of her amazing illustrations to show him. Her face lit into an excited, angelic smile when she looked up from her creation to see him approaching. Glancing at the table, he knew what he would see before his eyes registered the form on the page. Beside her pencil tip was a marvellous, artistically created woman with wide, detailed wings spanning almost the whole width of the paper. Her hands were clasped together over her heart. Black, flowing hair framed the precious face he knew so well. She appeared to be suspended in the air with a gentle wind current ruffling the white, silky gown. Her feet were bare and crossed over one another like in many paintings of Jesus Christ on the cross. She was extraordinary. Adrianna was an angel to him, but Dierdre drew her as an angel from heaven—a heaven he didn't even know existed or not. What it meant he was sure he knew, and he was convinced that this and many other facts about Adrianna would be revealed in time.

"Dierdre, that is an unusual but imaginative picture you have there. Do you know who you have drawn?" he asked, hoping to achieve a response of some kind.

She stared at his curious face, her smile never wavering, although a glint formed in her eyes that suggested that he was being patronizing and she thought he was an idiot for the question. "I'm sorry, Dierdre, that was rude. I know you know who that spectacular woman is, and I am betting you know a whole lot more than that too." She grinned wider at his words, revealing the intelligence beyond her years.

"I came to ask you if you would like to join Adrianna and me for a short walk to a very pretty place. She is waiting outside and would very much like to meet you." As the question came out of his mouth, he realized he had never seen or heard of Dierdre leaving the common room. Sadly, most of the time he did not think too much about her. She was always quiet and was therefore left to her own devices. This made him feel a guilt he didn't expect, and a sorrow he knew somehow she wouldn't want him to feel. His face pained, he glanced over to see her reaction to the invitation. She was standing next to him, looking up, eyes a sparkling sea green. Her pictures were stacked neatly on top of the desk in the corner, the newest one not on the top of the pile.

Holding hands, they walked out to meet with Anna, whose irritation had probably multiplied by now. She was standing by the blueberry bushes, stuffing dozens of the ripe, large berries into the pouch she had created by holding up the bottom of her top. A look of surprise replaced her previous look of concentration and mild annoyance at being kept waiting. A huge smile brightened her face and the area around her when she saw the darling girl standing next to Jonathan. She walked over, quickly offering her new-found, tasty treasures. Dierdre's smile matched Anna's in intensity and joy. They were going to be great friends, Jonathan felt very certain. Without a word spoken, Jonathan led them on a narrow gravel path to the left. The berry bushes lined the side of the path for miles, and he knew they led right to the meadow. He hadn't been here in quite some time, and didn't know why he hadn't thought of it earlier. He could have asked Dierdre here a dozen times in the past, he thought, his shame returning briefly.

Guilt brought on the awareness of the time he had been spending ignorant to the happenings around him. He had no idea if things had become worse for Shadow or any of the elders. He knew that it would be his fault if they had. After the meadow, he decided, he would tell Adrianna what he needed to and take whatever came his way. He was terrified of being wrong in his confidence, hurting Adrianna, and failing in everyone's eyes. Time would show if he was to be worthy of redemption and if he was worthy of Adrianna.

The excitement radiating from the two girls was almost tangible when they saw the field come into view at the bottom of the rather steep incline they had just reached the top of. A few white, fluffy clouds had gathered in the sky, making nameless forms to engage the imagination. The meadow was very large, framed by evergreens in a perfect circle. The snow-kissed mountains could be seen behind the tall treetops. It was thick with butterfly weed, prairie aster, Shasta daisy, and white yarrow. Sparrows, blue jays, and the rare mourning dove could be seen in the distance, their songs and calls heard from all directions.

The bundle of blueberries now eaten, Adrianna and Dierdre left Jonathan behind as they ran, arms wide, into the soft grass. They pranced through the colors while insects jumped and flew to safety. They looked like angels flying with grace through the magical field. A chestnut-brown rabbit with her babies bolted from its hiding place and Anna laughed with glee. Dierdre waved Jonathan over and he ran with exaggeration, flailing his arms and legs, singing "The Sound of Music." They laughed heartily at his silliness. It was impossible not to laugh with them.

"Oh, Jonathan, how is this even possible, to have so many beautiful places to visit every day?"

"Well, as I started explaining to you before, anyone can visit all of the places in this world; however, you don't have to walk great distances either." He turned his focus towards Dierdre, who was paying very close attention to his every word; it made sense that no one had explained anything to her, as she did not ask questions and everyone seemed ignorantly afraid of the little girl who couldn't speak—the only child in the middle world. He continued in his soft, charismatic tone.

"It is believed that most of the places you can visit were created by many of the first ones in the middle world. They simply thought up a place from their living past and it was created, not unlike Adrianna's ability to create the furnishings and the room she found herself in. However, not everything can simply be produced. It is my experience that there must be a need for something to be formed. That need may only be for the one person who thought of it, or for many to derive pleasure from, like the meadow. A few of the places the middle-worlder's visit are on earth, in the living world, unknown to the living, of course. Though there are also many places and rooms that have no connection to the mortal world, like the common room, for instance. No one knows exactly how it was all formed, and many learned not to question the fortune of being able to visit and see the living while having many places to be away from that world . . . ." He broke off, thinking that if he kept going he might lose their attention, although he had explained this and much more to Adrianna already.

Smiling, he said, "ok, holy peanut butter toast, that was a lot to take in. Just enjoy it here, girls; someone gave us this place."

Nodding in unison, they inwardly appreciated the gift of the meadow. Adrianna looked at Jonathan as if she wanted to say something but didn't have the courage or the privacy. He wished they could be alone, as they had been for so much of the previous two months. He had learned about her favourites, like the colour purple and the old black-and-white, classic movies she loved, from even before his time; that she loved fishing but hated fish; loved to dance and sing, though shyness prevented her doing these in public; and that secretly, for a long time, she had wished she had been born into a rich family in the United States, where it was easier to become a child actress and advance into a complete acting career.

He had told her that he never attended much school past grade nine. He helped his father run the store on the docks, while his mother cared for his two brothers and one little sister. Sharing his love for books of any kind brought a blush to his cheeks, and he felt a shock through his body when she showed delight in his words. The stories of street hockey and reading to his mom and siblings were a joy to share with her. He had never told anyone much of his past, specifically about his mother.

Growing up as a fisherman's son wasn't a glorious tale. Explaining how you could count on your fingers how many hours you spent bonding with your father was flat out sad. He wasn't embarrassed over the life he was brought up in, though discussing it with anyone was never the first thought on his mind. The events that had taken place near the end of his mortal life had never been discussed.

Watching her astonishment, curiosity, exhilaration, and animated desire for more elicited feelings of such delight in Jonathan, that he once again shoved aside any thoughts of earlier conversations, and the urgent issues at hand, until Juniper decided to visit them. Juniper smiled as she approached them. Jonathan was wary, Adrianna was relaxed and appeared excited to meet, yet again, someone new. Shockingly, Dierdre looked apprehensive. She held out her hand first to Adrianna, who took it with a gentle clasp then hugged her. Juniper, startled, glanced at Jonathan

"Everyone was right when they said this girl had enough love to go around. How are you three enjoying the meadow?"

Jonathan relaxed at her friendly tone and the obvious like Anna had for her. He liked Juniper as well, but the worry of jealousy crept into his mind. She had much to overcome and many believed she never would.

"The meadow is beautiful, Juniper, as you well know. It was you who first brought me here so long ago, was it not?" They both knew it was Juniper who had introduced Jonathan to the field. It was a short time after he had arrived and he was still having trouble letting go of his life and death. The meadow had been a place of reflection, a place to be alone, quiet, and serene. He had loved it almost as much as his park bench, but not quite. Wondering why he had not visited it again until now was a mystery to him, but now he would visit it as a place to remember seeing Adrianna dance among the colors and butterflies.

Juniper turned back to stare at Adrianna. She looked almost sad, but also relieved. He could see a million thoughts racing through her mind, some of the past, some of the not-too-distant future; he wondered which would take precedence when she decided to speak. She surprised him by speaking to Adrianna.

"You are very beautiful, Adrianna. I can see why everyone is so taken with you. I was in love once, to a man who was a lot like you. Kind, giving, a pure heart, although he wasn't as strong, nor as young in mind or heart as you. He is lost to me now. Maybe I will find a way to see him again. Maybe he won't be lost forever." Her eyes had left Anna and trailed off into a vision only known to her. Jonathan imagined she was seeing her husband in their finest moments, when their love could have conquered everything. Not wanting her to continue and risk upsetting Anna, he put his hand on her arm.

"Juniper, perhaps this isn't the time to go down that road. Anna hasn't been told your story, though I am sure a day will come when you can share it with her. Today, however, is not that day. We will be heading to the common room now. I have much to discuss with her."

Hoping this would satisfy her, and not raise too many questions with Anna, he held out his hand to Anna and Dierdre so that they might make their short journey home. Juniper seemed slightly annoyed by the shortness of their conversation. Recovering quickly, she began rambling about things that she must have known Anna had already been told.

"Travelling is an interesting action in this world, isn't it? You cannot exactly think yourself anywhere you want to be. If the place is near the place you were and has been previously thought of, then you can just want to be there and there you are. I love to think from place to place, just to see how fast I can do it, and then, like Jonathan, I will stop for a while and get lost in my thoughts. Middle-worlder's lose the ability to create new places shortly after adjusting to the new existence. I created a room or two and the chapel where my husband and I were married, although no one seems to visit it but me."

By the time Juniper had finished speaking, saying most of what Jonathan had told them earlier, they had arrived. Dierdre waited behind, staring at Anna. Jonathan noticed this but decided to walk Juniper in to take her away from the girls. Although Anna didn't seem annoyed by her ramblings, Dierdre was agitated and was even biting her nails. The door shut behind them with a loud click, and it was moments before either of them moved. Adrianna turned to Dierdre and crouched on her knees in front of her. She gathered the small, fragile child in her arms and spoke softly in her delicate ear.

"Thank you for coming with us to that wonderful place. It was lovely to meet you, and I hope to spend more time with you." Just as the words had escaped her mouth, she felt a bitter-cold slap against her clothes and bare skin. Even the sky seemed to darken. It was as if a heavy, icy blanket had been dropped over them, threatening to freeze their happiness away.

As suddenly as the icy feeling had engulfed her, she lost her surroundings. She could see nothing, although she still had the comforting feeling of Dierdre in her arms.

The blackness scared her and brought up memories that she wanted to forget. Suddenly, a new vision formed in front of her eyes, one she knew did not belong. From a distance, she could see a crack of light coming from a door, in a dark room. The hallway outside the door let a slight glow into the room. A baby's heart-wrenching wail could be heard close by. Suddenly, the crying stopped, followed by a slamming door and heavy footsteps in the hall.

Fear welled up inside of Anna as the movement in the hall came closer to the dark room she was apparently in. She felt Dierdre shake in her arms, instantly realizing she was seeing this too. A large-framed man entered the room abruptly and quickly made his way through the clothes and stuffed animals piled everywhere. The light from the now-wide-open door revealed a small frame hiding under a blanket on a mattress lying on the floor.

The man grabbed the blanket and threw it back, uncovering a dark-haired little girl in pink PJ's with white bunnies. Grabbing her shoulder, he spun her toward him. Adrianna jumped, violently shivered, and held her breath when she recognized Dierdre as the scared little girl. The man leaned close and growled all too loud in the tiny little ear under his angry breath.

"Keep your mouth shut or I will shut it for you."

Jonathan's voice broke through the image and cleared the air in an instant. Adrianna would have hugged him if she weren't glued to the spot in shock and pain for Dierdre. His voice was her saviour once again. The image still lingered in her mind, but the atmosphere had returned to its previous state. Dierdre, however, was still trembling in her also-shaky arms.

"Hey, you two, this isn't the last time we will see the meadow or each other." His voice was filled with curiosity and underlined with worry at the looks on each of their terrified faces. He didn't ask what had just transpired, though the current in the air told him something sure had. He hoped Adrianna would offer an explanation, sooner rather than later. The girls released their hold on one another and Anna stood with apparent weakness. They both appeared to have been through some traumatic event that Jonathan had missed by just minutes.

"Let's go in, you two lovely ladies, before someone thinks I have taken each of you captive to have you all to myself." He forced humour into his voice and smiled when he saw the slight grin on Dierdre's face. Anna, however, did not seem ready to be overcome by his charm. They resumed holding hands and entered the common room together. Many smiling faces met their entrance, while some were shocked that the sweet mute girl had actually been with them. No one spoke, though their thoughts were evident.

Jonathan advised the two girls that it was a great time to get some rest after their adventure, and they each nodded in agreement. He said goodbye to Dierdre and thanked her for joining them, promising there would be another journey soon. Excusing himself from Anna for a moment to look around for Valerie, he noticed Dierdre looking at Anna with deliberation. As he walked away, Dierdre motioned for Anna to lean over towards her. Jonathan continued walking, not waiting to see Dierdre try to speak to Anna.

After leaving Adrianna pre-occupied and hopefully resting, and not having the courage to ask her about her departure from Dierdre, Jonathan fled to his bench in the park. Although it was well into night on the living earth, he knew he was right to come here. The turmoil swirling in his heart and head eased just slightly while he sat staring at the stone birdbath to his right. It was illuminated by the solar-powered lights lining the walkway through and around the park. A light blanket of snow covered everything around him, though he did not feel the cold. He found such solace in this place and he would be able to reflect, although he had to admit it was lonelier here without Adrianna.

He remembered the first time he had come back to the living world after he had died. He hadn't been in the middle world for long, and Shadow had been his only friend. Not well versed in the rules and limitations he was bound by, he had wanted to check on the girl he had been dating, so he had simply "transported" himself to her Shakespeare class at Bay View College. He had sat at the empty desk two seats away from her as she wrote a synopsis of Macbeth in her pretty handwriting and chatted lightly with her friends. She had cried for the first few days after he had gone missing, cried harder when his body was found, and then she had moved on. Jonathan was furious when she had started dating his best friend, Mackenzie. It was 1959, a year when eight of the planets aligned for the first time in over four hundred years, John Lennon got married, and Jonathan had just won a '59 Chevy Impala, with a standard 235 engine, in a bet on a fight.

Other memories of the last year of his life, particularly the last few months, invaded his mind like a relentless disease. Even his favourite place could not keep the sadness hidden deep in his mind where he had stuffed it so long ago. He gazed around, trying desperately to find a pleasant thought to stick to, just as he had advised Adrianna. Yes, Adrianna. So beautiful, so remarkably soft, tender, intelligent, and forgiving. But would she forgive him when all was said and done? Would she be able to look past the fact that he had brought her here to use her amazing attributes to find a way to keep their middle world, and would she see that he really did have feelings for her? He did have feelings for her. More than he wanted to admit, more than he should have.

It was frowned upon to become romantically or intimately involved in the afterlife. The elders believed that the living world was where a person found love and that they should be content in that love even after they died. They also believed that the volatile nature that many relationships have would upset the equilibrium of their world. When two people fall in love, they grow together and need constant change and learning to keep their love flourishing. There is a never-ending need to progress and to experience new things. This could not be achieved in the static existence they all knew and had adapted to. Some of the more spiritual elders, who sat and contemplated everything all day long, believed that love and lust in their peaceful continuation would invite interference from the "other forces" of a darker nature. Maybe those forces had already interfered, and that was why Shadow really was now just a shadow.

Jonathan tried to remember the quotations from the Bible that the elders preached at many of their open sermons. They said that Jesus stated there would be no marriage among the dead or the angels in Heaven, and something about it being too big of a commitment for eternity. He had laughed at that, as his longest relationship had been six months. It was different now, though. Being faced with an eternity that up until recently had been concrete gave you a completely different perspective. In life, there is so much to experience, so many people to meet, and many that make you feel diverse feelings.

Commitment was a rare gift, not often given or received. With no shortage of time, it seemed more pleasing to want to give yourself to one person indefinitely, especially if that person is your soul mate. He had not thought of the concept of soul mates in a long time, but many said it was possible to be spiritually tied to one person.

The sensation he felt from the last kiss he shared with Adrianna flooded in again and nearly drowned him. He could still feel her lips on his, her hot breath coming fast and sweet. When she had grazed his chest with her fingers, he wanted to lift her off her feet and cradle her against him, to feel every limb, caress every inch of her skin and lose himself within her. Shocked at his own thoughts, Jonathan got up from the park bench and started walking the trail he knew so well. He had a serious problem here. He needed solutions.

Jonathan decided he should ask for advice—not something he revelled in doing, but necessity prompted some desperate measures. Asking Juniper would serve no purpose, as she was biased. She had died in a skiing accident when a few of her fellow skiers, including her millionaire husband, had caused an avalanche. Her husband had survived. She had visited him each day on earth and had tried every trick in the book to make sure she was not forgotten; she had even enlisted Jonathan's help, since he could speak to people and manipulate things in the living world. Sadly, she had pushed too hard.

Despite warnings from the elders, Valerie, and even Jonathan himself, she began sabotaging his relationships with others, leaving him notes to remind him of her, and somehow (no one ever figured out how) she learned how to invade his dreams. He had been unable to overcome his grief at losing her in such a tragic way, which he had a part in causing. Her husband grew depressed and eventually suicidal. He committed suicide after waking from one of his persistent nightmares and was lost to her forever. Her one true love now sits on the cold stone floor in a segregated room in the middle world, with many others who have taken their own lives, who remember nothing of who they are, who they were, or who they once loved.

No, Juniper was out. Shadow had way too many other, more pressing things to worry about than Jonathan's finally admitting that he might not be as sure about his success as he had promised. The other elders had no time for him, nor would he want their time if they did. The hundreds of others that shared their reality were scattered everywhere and sometimes difficult to locate, and he took no comfort in revealing his guarded heart to any of them. He decided, after some debate, to seek out Valerie. She was harsh but honest, and didn't have a broken heart.

To find Valerie, he decided to try the frequently visited common room. This was where anyone went with questions that needed answering, or just to communicate with someone. It was also the place where most newbies ended up and where she would most likely be. Not wanting to stay too long in the physical forum room to listen to the gossip, chatter, and sob stories, he briefly looked around for Valerie. Dierdre was standing just two feet away when he saw her. She seemed to have recovered from earlier, as she was smiling pleasantly once again. She abruptly pointed over to the far-right side of the room. There stood Valerie, who was talking to two new car-accident victims.

"Thank you, Dierdre," He said with a smile, walking over, and kissing the top of her inconceivably intuitive head. Following her direction, he started towards Valerie and cringed back momentarily when he saw the expression on her face as she noticed him advancing. Maybe it wasn't such a good idea to talk to her, he thought as he resumed his slow approach.

Valerie broke off her conversation with the drunk-driving fatalities and met him in the middle of the room.

"It's nice to see you, Jonathan. It's been a while, though you came quicker than I thought you would. I am assuming she didn't slap you and banish you from her sight," she remarked with a humorous snarl. He knew she was trying to be funny, with an underlying seriousness. It really did make her angry that she could not "see" everything to do with Adrianna. It was actually nice to see Valerie anyway, as he had not had even a moment to speak with her since Adrianna's arrival. They had crossed paths, of course, and he was quite certain that she had been told of every trip they had taken and every over-heard conversation, but Valerie had stayed back and allowed Jonathan to acquaint Anna with her new world.

"No, Valerie, she jumped into my arms, smothered me with kisses, and told me she had been waiting for me for an eternity," he retorted quickly, the memory of her warm, wet lips intruding on his thoughts. He prayed Valerie hadn't seen that. Sarcasm out of the way, they walked towards the only empty corner to continue their conversation.

"Valerie, I came for advice, which in all honesty is not easy for me to do. Can you be open-minded and civil please?" Swallowing his pride, he spoke with sincerity and desire to soften her up, and because he did not think he would be able to keep his anger in check if she kept up with her humour-coated scorn.

"All right, Jonathan, speak."

"I have spent quite a lot of time with Adrianna, as you probably know." Even just her name sent a shiver of pleasure down his spine. "Our first meeting was . . . well, wonderful. She was . . . well, quite taken aback at first over some of the things I told her, although I did not give her too much information." He was trying very hard not to think of her sensual embrace or the feeling of "home-coming" he had while in her arms.

"Since our first meeting, we have visited many places and talked a lot. I have not told her about why she is here, but I have explained much of what happens here and how we get around, socialize, rest, and just exist. She has had a lot of questions, not all of which I wanted to or even could answer." Jonathan remembered the conversation she had tried to start about angels and how he could not explain anything to her.

"That sounds like it went over quite well, Jonathan. I am glad for both of you. She must be remarkably strong and intelligent to be able to adapt to that much information so soon. Though I am failing to see what you need advice about." The genuine surprise she felt at Anna's poise reverberated through her words.

"It's like this, Valerie: Anna is tough, but I have, through speaking to her, come to the realization that she is not going to handle the . . . let's say . . . publicity well. She does not believe she is exceptional in any way and she does not want to be. She would like to fly under the radar, as she thinks she always has. I told her a few things about her, ummm, some of her more commendable qualities and she . . . well, became very upset. She is very sensitive and although she has come further than anyone expected her to in such a short time, I am afraid if I tell her anything too quickly it is going to be too much for her. But I have not seen Shadow since meeting Adrianna, and I am worried that he is still fading fast, so I feel like I am out of time here." The words tumbled like a rockslide, building fear, and tearing down his confidence as they went. She considered his words and gained insight immediately to the subtext.

"You are afraid to hurt her with what is expected of her, with informing her that she is imperative to our future. You do not want to fail Shadow or any of the council because you feel this is your only key to redemption for all you have done and not done in your living past and here. You want to prove you are not just unique in your ability to communicate with the living and your intuitiveness about the thoughts and feelings of some. And, of course, you have fallen even more in love with her than you thought was possible."

This last comment was like a blow to his very un-beating heart. Hearing the words spoken clearly, concisely, and full of an unacknowledged truth was more than he could bear. Looking around the busy room, paranoia swept over him. 'Had anyone heard? Would they care if they had?' Questions plagued his mind to distract him from the truth that Valerie had shot through his walls, his impenetrable fort of defence. He could not deny this anymore, knowing now that of all the unanswered questions he rightly had about how to advance Adrianna safely, how to resurrect the shelter of their world was not the real query; it was how to succeed despite his emergent love for sweet Adrianna. Valerie knew this. She knew that his quest had changed without his awareness.

"Yes, Valerie, as usual your perception is greater than the force of evil, and you have, inarguably, hit the nail on the head," he admitted ruefully. There really wasn't any point in denying it to her or himself any longer.

"Jonathan, this truly is a difficult situation we have here. We need Adrianna to complete her awareness of our world, in a timely but safe manner. We need for her to meet the crucial people and search her soul for the answer to our plight. Your love for her cannot interfere. You must act according to what is right to sustain our future. I need not tell you that your loving her, and her potentially loving you in return, will not be looked upon highly. It is not acceptable. However, you may be surprised to hear that it is by providence that you two are in love. I do not know why, as it has never happened here before, that I know of. But I had a vision of Adrianna and you. I will not explain more than this: you and she are destined to be together in this middle world, and you will both leave this world together. Be warned, Jonathan: this is the outcome that fate has determined. It is not absolute. It is not set in stone. It can be changed if a mistake is made to disrupt the order of events. I would be sad to see that happen, as I believe there is more change for both of you. You are both more special than any of us know."

Each word came from Valerie's mouth in wisps of the only air he could breathe; he took them in, inwardly gasping, releasing the stranglehold on his throat with each breath. These were not the words he expected. They were supposed to be in love! They were destined to break one of the biggest cardinal rules in their middle world. It didn't seem possible to believe such a vision. A terrifying vision. Jonathan felt trapped by the thought that he would destroy his world, his friends, and his own destiny if he should fail.

He did love Adrianna, this much he did know, but he also knew he was nowhere near good enough a person, in the past or in the present, to be destined to be with her. That could not be true. Anna was so good, true, and pure, and he was so tainted, and after what he had done to her, to her very existence . . . If he had the option of being around her each day for the rest of eternity (which he secretly hoped he would), he still would not nearly be worthy of her loving only him. Valerie's words continued to circle around in his head until he felt dizzy. She wore a pained smile full of empathy that Jonathan did not know she possessed.

"I thank you, Valerie. Although I do not believe your vision, I know that is what you see and you believe it to be true. I will talk to Adrianna and explain what is happening here and why she is important to us." He turned to leave and do some more reflecting over what he had learned, when Valerie placed her hand on his shoulder and wheeled him around with some force.

"Don't be an idiot, Jonathan. What I see is not of my opinion, whether I wish to feel happiness or anger about it. I have the gift of the future, but it is very limited. I can see a lot of your future and even your past." She glared at him as those last words left her mouth, and he knew that she had seen more than he wanted her to know.

"I have told you what IS destined to happen, and if there is a "god" who does decide our fates, then this vision is his vision. It is not for you, nor I, to determine if it is best or not. It just is. Obviously, you and Adrianna are meant to be together, for some higher purpose of good, whether you think, after what you have done to her, you are worthy of her love or not. Maybe you're not. But something or someone thinks you are. Now of course, if something happens to change things, then the future will change as well. I suggest you get your head out of your arse and realize what this all means. You must talk to her, you must protect her and you must save us. Oh, by the way, Shadow is looking for you."

This time she turned to leave, though he did not try to stop her.

8

Revelations and loss

How one was supposed to rest with so many crazy things happening all the time I had no idea. The argument with Jonathan, although still in my mind, had been replaced momentarily by the wonderful visit to the meadow, the horrible vision I had seen with Dierdre, and what it all could possibly mean. She, without a doubt, was the little girl in that dark room; I was in her room too, replaying her memory with her. She had appeared to be as startled as I was by the surfacing of her recollection, which seemed to have been prompted by our touch.

I had no doubt that she was as shaken by the vision as I was, as her little body had trembled when I was hugging her. It must have been a horrible night for her, and I intended to find out what that awful man had done to her. Oh my god—did he kill her? Was that why she was here? But why hadn't she moved on, as Jonathan said? I thought my parents were bad, but that guy was just plain evil.

Wishing I could just ask her what had happened made me realize that if Dierdre had wanted to tell someone about her living past, she would have tried some way to communicate, maybe sign language. She had tried to tell me she was sorry for the vision she gave me when we hugged for the second time. Maybe she just didn't want to talk to anyone. Dierdre was awfully alone, that was for sure, and I wanted to make sure she didn't have to stay that way.

Jonathan obviously adored her too, and she him. When he had told the story of meeting her and the times they had been together afterwards, he had a softness to his face she had not seen before. He took on the brotherly love role as if he were made for it. He no doubt had a lot of love for the siblings he had briefly mentioned on one of their excursions. I wished I had seen him playing with a sister or brother; he must have been a great role model for them. I thought of Steph, keeping the pain at bay, hoping I had been a good example for her as well—until I died, that is.

I wasn't getting any rest at all with all these thoughts plaguing my head: Jonathan, Dierdre, my new life, my old life. So many distractions from what I knew I should be doing. Well, I knew a lost cause when I saw one, and this was one. I was not going to rest yet, despite my growing mental fatigue. I was going to find Jonathan. We had much to discuss. I wanted his agreement that we (oh man, there I go again with the "we" crap) should be asking Dierdre to go on more of our outings, or just to hang out with us. I also wanted him to be honest with me about whatever it was he was keeping from me.

The argument we had had made me realize that it was time for each of us to come clean about our intentions: for my part, that I wanted to learn how to communicate with Willie and, for his, whatever secret he was keeping so close to his heart. No matter how afraid I was, it must be done.

Looking around the room I did not spend much time in, I discovered with amusement that I must have added a few trimmings to it, without realizing. Well, more than a few. In the corner of the now-light-mauve-coloured room (one of my favourite colors), I noticed an old-looking, dark-stained bookshelf, completely stocked with books whose contents I had no idea of. Beside that stood the writing desk I had had all along, which was now adorned with a high-powered LED reading lamp, a box of pens and pencils, a black leather-bound folder that looked to be full of paper, and the writing tablet I had barely used. In front of the desk was a black ergonomic chair (always wanted one of those) with a darker-mauve sweater laid across the backrest.

Glancing around with much more attention now, I realized that my comfortable white leather chair, which I had moments ago been sitting in, had transformed into a matching loveseat with a small all-glass table in front of it. On the table was a white lace doily like my grandma used to make (she had also tried to teach me the tedious craft, with little success—I'm all thumbs). There were still no windows, which creased the frown I was wearing, and I felt my lips purse unconsciously at the irritation I was suddenly feeling.

Either all this stuff had just materialized in the moment I decided to leave the room, or I was less observant than I originally thought (I could be pretty dumb at times). Why it had all just shown up, I had no idea, but I was sure that it had a purpose along with everything else. So, I did what seemed to be the right thing to do: I investigated my stuffed bookshelf. It had a faint odour as if it had been in someone's attic or garage for years. The books, as well, held the musty smell and appeared to be old. Some were cloth bound and others were regular paperbacks and jacketed hardcovers. I had always been an avid reader as a child and teen. It was only when I started to feel the need to be gone more than at home that my reading began to take second place to most other things in the evenings. At home, my collections included everything ever written by Stephen King, Dean Koontz, and John Saul, to name but a few. Of course, the classics were there as well: my personal favourite, *Gone with the Wind*; Charles Dickens's works; *Moby Dick*; *Treasure Island*; an Alcott book or two; and another close, second favourite, *Alice in Wonderland* (HA, where is my Cheshire cat and tea party?).

I scanned the titles and was flabbergasted when I realized what my shelf was so ingeniously stocked with: *The Joy of Signing* (I guess I was going to learn after all), the mystery surrounding the Archangels, the Bible, and *Relaxation for Dummies*. I couldn't help but laugh aloud as I realized that many of the questions I had, and the goals I had to achieve, could be answered, and accomplished with what was in front of me. Boy, did I have a lot of reading to do, and plenty of time to do it in. Scanning on, I saw a few titles that meant nothing to me and a few that piqued my interest. I read these aloud, as if it would help to decide their relevance: "*The Power of Intuition, Understanding Trauma, Tapping into your Heart, Love Explained.*" Hearing my own voice read the names of the books I was sure to read enlightened me to the fact that they all must have some important significance in my new existence; just how urgent it all was eluded me. I felt a sudden onslaught of apprehension when I thought of all the information that I needed to absorb; I wasn't sure why this caused such a sense of dread, like I was running out of time. I had loads and loads of time. I was dead, for heaven's sake, what was I worried about?

If Jonathan was right, then I would have eternity to learn how to talk with Dierdre, how to hone in on my intuition, and whatever other enlightenment these chapters could provide. Maybe the book about trauma was to help Dierdre with what had obviously happened to her in her past, though I would have to learn to converse with her before I could offer any friendly counselling, and I was sure that my own past would offer a certain empathy with her undoubtedly much more terrifying and short childhood.

Whatever the rationale for these volumes' showing up, I was ready to take it on with a renewed urgency and energy that I didn't know I could possess.

Grabbing the first book off the shelf, I quickly searched through the pages to figure out how to say "hello" and "goodbye" in sign language. The actions seemed simple enough, though remembering it all would be the challenging part. If I wanted to find out what had happened to Dierdre, I would need to learn much more than greetings and farewells. Flipping through the pages, I realized it was going to be difficult to learn how to ask "Who was that evil man who scared you at night?" and of course, all this time I am assuming that Dierdre would be able to interpret sign language and respond the same way. She could say "I'm sorry," but that didn't mean she had a wide visual vocabulary.

I was pretty sure she was only about five years old. What could a child learn to sign in five years, if she had even been mute that long? Jonathan had said she could hum, which indicated a working voice box, didn't it? And it was apparent she could hear, unless she was a genius at lip reading. I really didn't know anything about her at all besides the fact that she was a very sweet little girl who seemed amazingly wise beyond her years, that she had been through a bad childhood, and that something terrible had happened to her. Oh, it was all so confusing already.

I felt some of my earlier determination fade and became restless with the open book. Closing the signing book, I grabbed another few books off the shelves. Settling in my comfy love seat (I missed my chair and wondered if Jonathan had anything to do with its replacement), I spent the next hour skimming through the pages, reading some of the more intriguing chapters and taking in as much as possible. After I had absorbed as much as I could saturate my brain with, I underlined some key points to go back to, and dog-eared at least ten pages in each of the three books I had skimmed. I was finished and ready—well, as ready as I could be.

I decided that the only way to put my new knowledge to the test was to go and find Dierdre. I would bring my book with me and maybe between the two of us we could come to a plan on how to proceed.

I left my room again, for the first time alone, and found my way to the common room, where I hoped to find Dierdre and not Jonathan. I didn't want to see him yet, as I had my mind set on seeing the amazing little girl I had formed such a connection with. Entering the room, I immediately felt very self-conscious, like I had a huge zit on my nose (I was so happy dead people couldn't get zits). Most of the occupants turned to stare at me and conversations ceased. I wasn't the center of attention but I still felt very noticed. Just about everyone I shyly made eye contact with smiled in response, which made me feel better. A few of the ones that Jonathan had pointed out before glared at me with an intent I could not fathom. I seemed to be a very interesting specimen to many, though I had no idea why.

Dierdre was there, thankfully (there might be a god after all), sitting by herself at the little table in the right-hand corner of the largest room in the middle world. It was brightly lit with fluorescent lighting and furnished for as many as possible. Metal folding chairs were open everywhere, as were hospital waiting-room chairs, a few modern-looking couches with matching coffee and side tables. There were books, magazines, and writing tablets scattered on the tables and in little side bins. The large room was not decorated, probably to keep it basic and not offensive to any spiritual or color preference. I laughed at my own synopsis of the plain room. There were a few open windows on the white walls, without so much as a valance or blind to cover them. It seemed that the sun never directly shone through the openings, though it provided a warm glow that was adequate for comfort.

I approached Dierdre slowly, as she seemed to be concentrating very hard on the picture in front of her. Jonathan had mentioned that she liked to draw and was very good at it, but he never told me anything about her pictures. From the side of her table I could see a figure on the paper; as I got closer I could see it was a man kneeling on one knee, his elbow resting on his thigh, one side of his head in his hand, fingers through the long, dark hair. It looked exactly like Jonathan! To my surprise, he seemed to be crying, as there was a pained expression on the side of his face you could see and what looked like a tear on his cheek. In the distance beyond his sobbing body was the back view of a female, who appeared to be floating away. She had long black hair, was wearing a long white dress and no shoes, and she had wings. In her hand was the hand of a small child. A little girl.

Two things shocked me at that moment: the drawing was better than any I had ever seen in my whole life (and she was five), and it was of Jonathan, Dierdre, and me. I wondered if she had more like this and if anyone else knew about them. She glanced up at me and smiled. At that moment, it seemed that the room went quiet and there was only me and her staring at each other. I glanced down at her picture and her smile faded to almost a frown. I wasn't sure if it was because she didn't like the picture, or if it was the content that made her sad. Either way I wanted to tell her how amazing her drawing was.

"Dierdre, I don't know if anyone has ever told you this, but that picture is fantastic. You are quite the artist." I didn't know how to express how positively sensational her ability was. How do you say that to a five-year-old? Better yet, how does a five-year-old draw like that? (I was lucky to pull off a stick figure.) But I had come here for a reason and I was worried we would be interrupted, so I threw all caution to the wind and opened my mouth to speak. One way or another, I wasn't leaving until I could find a way to talk with her.

"Dierdre, that is the three of us in your picture, isn't it?" I asked, hoping that starting with the obvious would ease the way. She nodded dramatically, still frowning.

"He is crying and we are walking away," I stated simply, not needing a response. She nodded her head anyway.

"You saw this vision in your head, like the vision you showed me?" I asked.

She shook her head no. I kept on.

"The vision you showed me happened to you, didn't it?" As she looked down at her picture, her head answered yes.

"You saw this image in your head," I pointed at her picture, "and you decided to put it on paper?" Looking back up at me, she smiled again and answered yes. Her hand pointed towards the floor under the table. There was a large blue folder with paper inside. I picked it up, hoping that was what she wanted me to do, and slowly opened the cover. Inside there were many other drawings by her: one of a beautiful valley, with deer in the distance and a cloudless blue sky; one of Valerie speaking to an old man who was wearing a long black cloak (ugly, like the monks used to wear); and a faded creation of Juniper talking to a large woman with curly hair. She looked like anyone's grandma, though looking closer I realized I recognized her from somewhere but had no idea where.

I kept flipping through the pages, seeing many people that I did not know and a few I was sure I had seen here. Near the back were two more pictures that caught my attention. One was of Jonathan and me walking hand in hand; there was no background to this picture, just the two of us, just as we had walked many times since I arrived. The last one was me. A way-more-beautiful me. I was depicted as an angel, with the wings and all. I was stunning. (Crud, why couldn't I have turned into this when I died?)

Closing the folder, I peered at her little face. She was beaming, as if she could sense how much I loved her pictures. How a person could look at them and not love them was well beyond me. I wondered if her gift had begun when she came here or if she was one of those undiscovered prodigy children. I wanted to ask her so many questions, even more than I had remaining for Jonathan (ok that might be a stretch), but I didn't know how. I simply handed her the book of sign language I had been holding between my knees and looked at her expectantly.

She studied the book cover, grinned and shook her head no. I guess she hadn't learned much sign language. As if on a spiritual cue, a man from across the room came over to greet us both. After we had shaken hands and introduced ourselves (although he already seemed to know Dierdre and surely had heard of me), he turned to me to speak.

"Adrianna, I have only been here for a short time; however, I noticed this little sweetie couldn't speak. I spent my fifty-four years on earth never being able to speak from a birth defect. I learned sign language to get me through. I tried to talk to Dierdre." He smiled kindly in her direction, trying not to be rude by talking about her as if she weren't there.

"And I realized that she does not know many signs. I have taught her a few but not many. She understands everything that is said even when there is no visual to lip read, so she is not deaf, and some have said she can hum a sad tune, so it isn't her vocal cords, like it was with me." He reached for her hand, obviously feeling bad for talking about her so much. She eagerly took it and held tight, indicating she was more than fine with the conversation she was not directly involved in.

"I believe something must have happened in her past to make her decide not to speak. However, why she chooses not to even now, when I can speak in the middle world, I don't know. Maybe you can help her with that. I can teach you some of the words that I have taught her, if you would like."

He asked gently. I was glad of his help but even more discouraged after he confirmed what I had feared. We had no open communication pathways. I shook my head and showed him the book she had set on her table.

"I will try a few things in here, Peter, and other than that I guess we will just have to come up with a language of our own." I was pleased to see Dierdre's face brighten up at my words. She looked positively giddy with the idea, and I wondered again why she chose not to talk if she clearly wanted to communicate so badly.

"Good luck, Adrianna and Dierdre. If I can help in any way, please let me know. I'm always here, in this room." He nodded towards each of us, his voice changing when he promised his location. He sounded acquiescent to the new place he now lived in, and I speculated that he was sad to be dead.

Dierdre and I returned to her table and I pulled up a chair. No one seemed to be at all surprised at our friendship, although quite a few onlookers were stealing glances our way. I was a little nervous about touching her, but I wanted to hold her hand or have some physical contact with her to comfort her and reassure her that I was her friend and only wanted to help. Moving my chair closer to her, careful not to get too close, I leaned my chin on my hands and looked at her precious face. She seemed to want to say something to me and her little brows furrowed together lightly.

"Dierdre, I want to help you and I know that you know I have questions for you. Is it okay if I ask some questions and you nod either way?" I had no idea how I would word what I needed to know in a yes-or-no-answer format but I wanted to try. Her smile was forgiving and kind, as if she knew the turmoil I was going through. She seemed to agree with my proposal.

"Ok, let me see here. Ummm . . ." I stammered stupidly, trying to choose my words cautiously.

"I have done some thinking and, ummm, reading . . . . I will say some of my assumptions and you can stop me if I get any wrong. So, wow, you are an incredible artist." She grinned and her tiny but sturdy body trembled with a silent laugh.

"You see visions in your head out of the blue and you draw them as you see them." I glanced at her to see if I was correct. She was frowning slightly, with a look that suggested I was close but just off the mark. I tried again.

"You have a feeling you want to draw, and the pictures comes to you as you create?" She nodded with enthusiasm, happy I had guessed so quickly. I knew the book interpretation would come in handy and I was right. I had read in the book of signing that some children liked to draw the thoughts that were in their head. (Yes, an A for effort for me.)

Elated at our quick success, I continued with energy.

"You see Jonathan and me together, and we are important somehow." (Whoa, where did that come from?) I had no idea where the certainty came from that prompted that statement from my mouth, but I had hit the mark again, as Dierdre nodded with agreement. A little frightened at what would flow out of my mouth next, I re-opened the dam of thought.

"You know things about the middle world that no one else knows but you don't know how to tell anyone—wait . . . no, there are others besides you who know." I studied her reaction closely; it guided my thoughts as much as my speech.

"You are afraid." This bothered me, greatly. Dierdre feared something here, in the land of no danger, no fear, no enemies, and no loss. What could she be afraid of? I had to find out, and I had an eerie feeling Jonathan had something to do with this.

"Something is changing and we are a part of it. You are going to help . . . me?" Flabbergasted at the path my train of thought was travelling, I still could not grasp how all this information was coming to me. It was as if my intuition was heightened to a higher degree than I knew was possible. Suddenly I felt a heat, a light pressure, and an overwhelming electric current flow through my thigh; I looked down to see Dierdre's hand on my right leg. That was it! It had to be. It was her. She was communicating through her touch to me, and if I opened my mind (and my big mouth), I would be able to speak her knowledge through me. I wasn't sure how I knew this to be true, but it was. This is insane: I have powers, she has powers, we are all dead, and now there is some whacked-out conspiracy happening that Dierdre, Jonathan, and I are supposed to figure out. (What a bloody soap opera!)

I asked the next question that came into my head, whether from her mind or my own I was not sure.

"Do you know why you are the only child here in the middle world?" Shaking her head "no" showed me that question was of my own making, as was the next.

"Can you speak but choose not to?" This time her answer surprised me: she nodded "yes." So, she could speak but would not. Why wouldn't she speak? She was scared to speak for a reason I could not understand. There was a root-cause to that, I was sure.

"You are here and haven't moved forward because you are imperative to a cause, one I don't know about and you don't fully realize either. We are supposed to meet, become friends, open this portal of communication, and work together. We are all special for a purpose." I was speaking aloud, my own thoughts and Dierdre's, though I felt like I was talking to myself, like I alone could hear my voice rationalize the craziest situation I could ever envision.

There was no point in stressing about it or even marvelling in the extraordinariness of the whole situation. Unable to comprehend the fear welling inside me, I decided to change the route of my quest, I decided to go down a different path.

"That man in our vision, he was your father?" Her smile faltered quickly, showing an obvious discontent with the change in topic, though she nodded in affirmation and did not remove her hand—her portal—which was the key to all understanding. Sensing she would allow me to continue with the topic, I did, though I was afraid to hurt her or make her remember anything that would shut her down.

"There was a child crying that night and your father went in the room with him?"

I looked at her for a sign that I was correct that it was a little boy crying—her brother, I assumed. She nodded once, pain streaking across her china face. Whatever had happened had been bad, bad enough to cause her to shake at the mere recollection. I did not know if I should go any further. Believing she would stop me if I went too far, I pressed on again.

"Your brother was crying and your father went into his room and settled him?" I felt stupid even thinking the evil father did anything as fatherly as "settling" a crying child, but I was nervous about jumping to conclusions worse than what had actually happened. Nor did I want Dierdre to tell me I was right in the atrocious images I had in my head.

The first sign of irritation came across her face, and I knew I had gone too far. I needed to be honest with my assumptions or elaborate on the subject I had touched on so ignorantly.

"I'm sorry, Dierdre, I don't want to imagine what happened with your brother, and I don't want to see you afraid or scared. I am a coward and you are so much stronger than I am. To have gone through what you did and to . . . ." A lady I had seen staring at us frequently over the last hour interrupted me in mid-sentence. She had a very pleasant face and curly, long brown hair. Standing just an inch or so shorter than me, with maybe twenty more pounds (it was awesome to think I could never get fat), she approached closer.

"I'm sorry to bother you, but I couldn't help but notice the charm you wear." She pointed at my chest and the mysterious amulet I wore.

"It is glowing. I mean, it has been glowing off and on. A few times it became quite bright, and, well . . . I wondered if you knew how it did that." She asked tentatively. Nervousness was splashed all over her face, and her voice shook. I felt conspicuous and was curious myself if what she was saying was true; it had never glowed before. The amulet was always just a jewellery piece I wore because I was sure it was given by Jonathan, even though that was never confirmed. I had no idea it was special too.

I couldn't see the charm, glowing or not, as it was on a short chain, but Dierdre was smiling knowingly as if she had seen what the woman spoke of. I didn't know what it meant and I was almost sure the little wonder child before me didn't either. I was sure hoping Jonathan did. I would be asking him soon.

"Ummm, thank you. I honestly don't know how it is doing that. I don't think it has even done it before," I answered the woman, with embarrassment at my lack of knowledge of my own adornment. The woman nodded quizzically and walked back to her group, and they all began to stare, smiling weakly. Turning my gaze from the group, I focused on Dierdre once again.

"Well, that was random. I apparently have a magic amulet, along with all this." I said.

"It seems quite fitting, doesn't it?" She smiled in response.

"Will you show me what happened to you in your past, Dierdre?" Confident she would know I was referring to the embrace that prompted the first and only vision she had endured with me, I asked the obvious. Hesitating slightly, she touched her right wrist just above her hand where a watch would be if she had one to wear.

I was confused at first, then it dawned on me she meant "time." She didn't know for sure but was hoping that in time she could show me what there was left to reveal. She was spent, tired, raw, and emotional from the discoveries I had made with her help. Her pallid color had paled even more, bringing a light bluish hue to the underneath of her glorious green eyes. Guilt spread through me as I realized I was the cause of this obvious exhaustion. Seeing this emotion on my face, she smiled and took my hand gently. No dark room appeared, no wicked man, no little crying brother, only the same heat and electrical current as before, when I felt her hand on my leg. She was ok. She was tired and worn down but she was pleased that we had had our "talk."

Triumph now showed, almost radiating from her little body—she was relieved to be able to converse with me through our strange connection. I knew exactly how she felt.

I bent down to kiss the top of her head and whisper "goodbye for now" in her ear. I assured her I would be back to see her soon, though I had a few things to do and we both needed rest, her more than me. I was invigorated by all that had transpired between us, and I wanted to see Jonathan. I wasn't sure how much or how little I should share with him. Telling him about my bond with Dierdre was pleasantly unavoidable but I wasn't ready to share the image from her past or the elucidation of how essential we all seemed to be to some bewildering plight I knew nothing about. The feeling that he knew more than I did hit me again like a slap, awakening a sureness inside me that we must talk, and soon.

Deciding to walk back to my room through the white, bare hallways was a quick decision. It was oddly comforting to hear the clicking noise on the painted cement floor of the flat, black skater shoes I had managed to materialize for myself earlier. (Ok, so it was a little extravagant but I am still a teenage chick, you know.) Though it was becoming more difficult to create things for myself, I was still able to bring forth a few items I wanted; then there was the appearance of the loaded bookshelf, which I hadn't intended to bring about.

Confusion was still my first emotion when taking in all the goings-on around me. Jonathan was a huge help, but even he did not have an answer for everything. As I thought of his name, I could hear his voice in my head. (How funny is that?)

No, wait—the voice was not in my head, it was coming from just down the hall, from my door. There was another voice as well that I did not recognize. If I approached I would be interrupting, and if I didn't I would be eavesdropping. Oh, so many dilemmas. Against my usual better judgement, I stayed my feet to catch the gist of the conversation. Listening closely, quieting my useless breath, I could hear clearly.

The stranger was speaking.

"None of those who have moved forward has ever come back, although there are those who believe it can be done. Theretofore, we cannot advise any of us what the best thing to do is. No one knows, Jonathan. No one even understands how we have all could be here, in between the living world and the place you go when you have moved on. I have seen the living world change from the world I grew up in to the world of technology, and in all that time, close to a thousand years, not one being, alive or dead, has had any answers for us, only theories. We only know just a little more than your friend Adrianna did a few days before she joined us. We do know that when some leave their lives, and their human form ceases to sustain life there, they travel here. Most do not." He continued without even enough of a pause to clear his throat.

"If you can accept what has happened to you, then your new world here is created. For those who do not accept, or those who have purposely caused their own deaths, there is no 'here'; there is no anywhere. They are lost to us and to the living world, and they cannot move forward. They collect in the one area that you have seen and sit in some form or another. Eventually they dissolve into nothing." Hearing the words that defined the aftermath of a suicide hurt me inside; the memory of my brother Kieran came rushing back so suddenly I had to cover my mouth with my hand to stifle the cry. So, he was lost to me, lost to everyone and everything. I had honestly suspected as much but hadn't wanted it confirmed aloud.

We were never that close, my brother and me. With him being almost ten years older than me and having left home when I was just six, our lives did not intermingle often. Even when there were family functions to attend, holidays to celebrate, and times when my mother would beg him to come visit, Kieran stayed away. He sent presents and cards, made a phone call here and there, but he never joined us. I hadn't even thought about him since arriving here; I guess I had forgotten him for the most part even before he died.

Pushing my brother's memory away, my mind immediately wandered to Willie, knowing I hadn't and would never forget him. Thankfully, that destructive thought process was cut short when Jonathan's voice rang in my ears. Although it was a beautiful sound, it was full of strong emotions I was taken aback by: belligerence, fear, and sorrow.

"Well, that's all fine and dandy for those souls, but what I am concerned about is us, and now, and what my next step is. The world we know will not be here much longer if someone cannot offer more input on how to proceed. I know your plan is to stay here and dissolve with the rest of the middle world, if you last that long, but others are not so sure. Some want to move forward now, giving up on the chance of saving what we know; most do not know this is happening and will not be given the choice until it's too late. However, we can't start a massive panic by telling everyone or there will be no one left to save. How did it come to this, Shadow? Why, after over a thousand years has someone or something decided our existence should not continue?" The anguish in his voice chilled my soul but did not have the effect his words did.

A sardonic dawning was blanketing my suddenly not-so-curious mind. This was what I was waiting for: the mystery, the truth—that the world I had just begun to belong to was dissolving because of some unknown adversary (did this mean I was going to die AGAIN?), and Jonathan was not going to tell anyone. Just how I played a part in this, I had yet to learn, and I definitely hoped I would be enlightened by the end of the conversation I was so appallingly snooping in on. I got my wish, and much more, when the conversation revived.

"I don't have all of the answers you wish me to have, young Jonathan. I did not want to tell you all of this—not now, if ever—but I see no way around it. I will make this short: any elaboration you want you will need to seek out on your own. There are others that share the universe with us, Jonathan, besides the mortals. There are angels. You know one quite well and have met another. They are the ones with intuition beyond what you or I could ever achieve. Their destiny, as it has always been, is to protect, to watch over, and to bring equilibrium to all of our worlds."

Hearing the words of the stranger, who I now knew was Shadow, I got the feeling that what was left to come would be more than I could handle. I listened intently, afraid to miss even a syllable.

"There are others, Jonathan, other angels that have deviated from the path of the Archangel. The history is long and complex, you may read on it. These other angels, which we elder's refer to as the dark angels, are more . . . let's say . . . callous in their actions within our worlds. One in particular: Lucifer wanted power, and the dark angel was fed upon that need. Normally they do not interfere in anything but a little messing around with the lives and deaths of the living-world mortals, but I am almost sure they or he may be behind what is happening to all of us now. I do not know how to stop what is already progressing so quickly, nor do I know of his motive. Someone has decided that our middle world is a hindrance to the living world, and they are acting to resolve this interference."

Shadow's speech paused, whether it was for impact or because he was finished his tale, I didn't know. I waited breathlessly for Jonathan to say something. I wanted to judge his reaction to the inconceivable information he (and of course I) had been given. I couldn't even comprehend what I had just heard. A short time ago my biggest worries were if my mom would stay sober, if my new stepdad would stick around, how my hair looked, and what color eye shadow I would wear on my next date. Now I was knee-deep into angels and the devil, fading thousand-year-old monks, falling head over heels for a gorgeous dead guy, and an adorable, psychic, vision-transmitting little artist.

"You should have told me all this before, Shadow. I never knew we were dealing with dark angels. This is more than I bargained for. How is Adrianna supposed to battle creatures from the beginning of time? It's impossible, no matter how special she is. I can't allow her to be in danger."

Shadow's voice rose to the point of almost yelling; I cringed at the sound and the anger within.

"Jonathan, Adrianna will cease to exist; she will die the eternal death with the rest of us. You have no decision here; you have no more power than me. You cannot protect her. She is our only hope. She has a power that none of us possesses. With that and the ones she is destined to unite with, she holds our existence in her hands. I think it is too late for even her, though, not that it is your fault—it was just the timing and the insistence of the ones who oppose us. You very well may have had her brought here before her time for nothing, my young friend, and for that I am sorry."

I felt as if my soul were leaving my body once again, there were pain and a detachment that I couldn't stand to bear. I felt a sound escape from my lips, knowing I was powerless to prevent it. I was not in control. This couldn't be happening. Words zoomed through my head like a rocket ship off its course, threatening to crash and destroy all its surroundings in one massive explosion. The pain in my head flashed and flared with each thought. I was brought here (did that mean I was not supposed to die? I was . . . what . . . murdered?), I was here to save this world from a dark angel, and I couldn't possibly be successful. (Of course, I couldn't.)

Jonathan knew enough about what was taking place to convince me I was special to him and wanted by him, for me to do as I was asked (the impossible). I felt used and cheated. Was everything he had said a ploy? A dream made for me to be manipulated into a position I couldn't possibly know the first thing about? Now I would be responsible for the eternal death of all my new friends, the boy I thought I could love, and the little angel-girl who had absorbed my heart. The terror, in the form of a scream of agony, ripped out of me with a fury I couldn't reign in.

Suddenly I felt hands on my skin, and whispers of panic, anguish, and urgency. My body was limp, no energy rushed in my dead limbs, no happiness lived in my soul, and no more sound escaped me. I felt myself being lifted, and then we were moving. I am dead; can a dead person actually faint? I could hear my own thoughts, feel pain and external touch, so this couldn't be fainting. Trying to hone in on the hushed voices around me, I could barely pick out a few scattered phrases: "She heard too much"; "What is happening?"—that was Jonathan, I would know that voice anywhere; "Adrianna is detaching, closing, she is in pain. I'm sorry, Jonathan, our hope is going with her" . . . "Will she come back, Shadow? Tell me, Jesus, will she come back?" The answer to his pleading question was lost to me under his words. Was I really lost? Detaching from myself to end up as a suicide, closed off from everything but my own tortured thoughts? I couldn't let that happen—I had to get a grip on myself.

Feeling myself being laid on my love seat by someone I hoped was Jonathan (his voice had been closer, I thought, but that could have been dreaming), I collected myself and opened my eyes. Staring straight up, the first face I saw was Shadow's, his thick, grey brows furrowed, his face stern and heavily lined. There was kindness and intelligence in his eyes. His long existence had given him a plethora of facts, experience, and opportunities to meet every type of person that was ever created. But this lengthy time, his complex history, and static way of life had made him cold, disconnected, and resigned. I believed he wanted to fade away into the nothing, just to cease to exist.

With a fear combined with misery, I suddenly realized I could not see Jonathan. Had he left me? As soon as the desolation surged through me again, I felt him lift his head off my chest, where it had been lying. There was complete and utter devastation in his eyes as he searched mine for signs of consciousness. I stared back, not knowing if I was really there or not.

"Adrianna, can you hear me? Please say something to me." Wanting to speak to him didn't dissolve the bonds on my heart and my throat. He had betrayed me. His secret went deeper than any I could have imagined him having. The care, concern, and love in his voice were forged from his need for me. Not the need I wanted him to have, not the desire I had believed he had. Our closeness, our bond, and feelings for each other were nothing more than an illusion, one I had held on to, believed in, and cherished. I should have known that there was no way I could have died some unknown death to end up with true friends and my own destined soul mate.

Closing my eyes, I still did not speak. I could not form the words filled with pain and wretchedness. I wanted to be alone, as I truly was—alone and deceived. He stiffened by my side and I knew he felt the rejection I was giving him. Finding my words at last, I spoke without feeling.

"You betrayed me; you used me, and nothing you said was real. I don't want to see you again. I had a life, friends, family; you took that and all my dreams away. I am not what you think. I am not special. I will never forgive you."

I felt a twinge of regret flare up when I caught the expression on his face. His look of devastation had been replaced with a too-quick acceptance that he was the worst person in existence. He had failed, lied, cheated, and lost. Hopelessness flew from him to pass over me. I felt his wasteland. How could I allow him to feel this way, despite what he had done? He had been there for me, helped and guided me, cared for, and soothed me, and that was undeniably real. Wasn't it?

No matter how guilty I felt I still could not let go of the internal torture that he had given me. I said nothing as I followed him with my eyes as he left the room. I was surprised when I saw Shadow had stayed, and wondered what words he would offer to further put a chill into my heart and soul.

"Adrianna, I apologise for how you had to hear the reality of our world. I, however, am glad you did. Jonathan's true love for you has blinded him in his quest. You are stronger than I originally thought, and I know even now, with your feelings of agony at the assumed deception, that you know in your soul you are here for a purpose, that you have the power needed to save the middle-world life we live. I cannot tell you how to feel or not to feel, but I can tell you that if you continue the path you are on now, you will be responsible for all our deaths. It is not for myself that I say these words, as I am lost already, but I do believe there may still be time to save the others, and ultimately Jonathan and yourself. Just do what you are here to do, Adrianna." He turned to leave and was gone before his last word had sunk in.

I knew he was right, no matter how harsh he was in his delivery; his words hurt but hit home. Whether I could forgive Jonathan for his deception was irrelevant: I did have a responsibility to the ones I had met, to do what I could with what influence I had been given.

"Thank you Shadow, I will try." I said aloud to the empty air. Though the pain was still plaguing my body and sorrow bewildering my thoughts, I knew I had to speak to Jonathan. That thought burned my mind and dug a hole through my chest; I felt cold, tired, and weak. I had no idea how long I lay still on the couch, motionless but writhing inwardly, re-playing the words Shadow and Jonathan had spoken without awareness of my company around the corner.

Shadow had put things into perspective. I knew I was being selfish in allowing myself to delve into my own depression, and not considering how Jonathan was dealing with what had transpired. Shadow had said he had been blinded by his true love for me. Was that possible? Did he love me? I thought I had felt it in his kisses, touch, and enthusiasm in my presence. I had believed he cared for me and wanted to be around me, in whatever way was acceptable and possible in this reality. Being absorbed in my own personal feelings about being dead, my past, and my potential for the future had distracted me from knowing I had fallen in love with the beautiful boy.

Remembering the words that had been spoken indicating that Jonathan had been the one to take me from my life re-ignited the fire inside me. How could he have done that—taken me from the world I knew, the only world that offered me the ability to grow, change, develop, and experience all I could learn to be? I could have become a nurse, stayed with Willie, Audrey, Nicole, and my sister. Even my unloving mother and her makeshift husband would be missing me and had suffered at my loss. Willie would not be destroying his life and endangering his soul, and I could have been there for him.

Mourning the never-going-to-happen-now's of my life brought tears to my eyes and down my flushed cheeks. I would never be a college girl, get married (although I had done the fall-in-love part here), experience the trials and tribulations, joys and unbeatable happiness's of being a mother, watching my children grow and succeed. Did he really take that away from me?

I couldn't fathom it all. There was no way HE could have been the one to do it. He must be blaming himself for knowing it would happen and allowing it, but it could not have been his hand. It wasn't him. Resigning myself to the facts was what I needed to do now, not bellyaching about the could-haves and should-haves. "Whys" would only serve to cause more pain and fan the flames still burning relentlessly inside of me.

Slowly, I lifted my fatigued body off the couch and made my way to the very hall where I had overheard the tale of the end of the world (as we knew it, anyway). Shuddering, I passed the spot where I had stood and then fell, passed the doorways that led to unknown rooms and places, passed the plain, barren walls and the entryway to the common room. He would not be there. He would want to be alone and in a place where he felt safe and comforted. He would be at the park. As soon as the thought entered my head, it felt wrong. Very wrong. He was not seeking comfort, he was encompassed in his self-torture, wanting to feel his own torture increase and stab his very soul. (Again, with this crazy intuition.)

Jonathan was hurting bad. Worse than I could imagine. I had to go to him, no matter what, but how to find him? Sensing things was relatively new to me, though I knew I could do it if I tried, just as I had sensed he was not at the park, and the confident internal voice told me how tormented he was at this moment. I concentrated on his face, the heavenly blue eyes, and the tantalizing smirk that played on his lips when he teased and laughed. All the vision did at first was to increase the burning inside me. Sad that his face would cause so much agony when once it caused so much want and need.

I couldn't think about it. I needed to focus on where he was and how to find him quick. I was not sure what bad things could happen while he was in the state he was, but I didn't want to find out. Nothing would surprise me now. But where else would he go? Closing my eyes, I felt his touch on my skin, his sweet breath on my face, on my lips. I did not know the place where he was, as I had not been there before, but I could almost feel his atmosphere. It was warm and dry, no wind but very bright. I saw two balls of light in the sky like two suns spreading their glow across the rolling dirt hills. There were huts made of stone and dried mud, no glass in the windows or doors. It was devoid of all life, including vegetation or animals. The place was desolate, like Jonathan's soul. I vaguely remembered a place like this—no, not a real place I had been, but a place in a movie. Someone else had re-created a movie scene, as with the lush valley and hills of the elfin village. This place was filled with bad memories of death and power.

Seeing the location in my mind, I felt myself being carried over the flat plains towards the only being (although not alive) that was anywhere near. I saw him sitting on the edge of a shallow canyon, his feet dangling over the edge of the fifty-foot drop. His shoulders were hunched like a ship whose sail had settled in the stale, calm air. He looked deflated and worn. My heart reached out for him before I was near enough to touch him with my hand. Sensing company, he looked in my direction. When I saw his drawn face, I couldn't help but cringe back a step. He smiled, but not with joy: it was a satiric smile full of anguish.

"Jonathan, I came to find you, to. . . ." I did not know what to say (as always, when I needed my words the most, my mind went blank).

"To talk to you."

"Talk to me, Adrianna? I don't deserve for you to talk to me. I deserve for you to be honest, to tell me to jump off this ledge in hopes it might silence me, even though we both know it wouldn't, just as we both know what will happen to all of us because of me. Because of what I have done to you and what I could not do. Don't talk to me, Adrianna; go back with the others and find a way to save them and yourself."

His words were full of self-hatred and remorse. Although I was slightly glad to know he felt guilty for what he had done, I was not about to stand by and allow him to wallow in self-pity. (Immaturity vs. wisdom on my part.)

"Jonathan, you must stop this. I don't understand everything that I heard; I don't know what I am supposed to do here, or how I of all people could help anyone, let alone save a whole world. But what I do know is that we are friends, and I truly believed we were becoming much more than that. I have never felt this way about anyone, or cared enough about someone to be happy about dying. You hurt me, and I am not sure how to forgive you but I want to try." Astonishment slapped me in the face when I heard the kindness and love in my voice for this boy. Was I really telling him how I felt, when part of me wanted to push him off the ledge and save him the trouble of jumping? (Wow, I was vindictive when I was angry.) Speaking without thought allowed the truth in my heart to flow; I at least owed him that, and hoped I would receive the same from him until the end. (Which didn't seem to be as far away as I thought it would be.) He didn't speak, and the silence was quite annoying, so I continued. (I couldn't stop once I started. Bad trait.)

"I do not feel special, Jonathan, but somehow I am able to do things here that I never thought anyone could do or should be able to do. Maybe it is too late, maybe I am not nearly as special as everyone thinks, but with what you can do combined with me, then isn't there a chance we can try to save this world and all those innocent souls?" I hadn't felt the tears streaming down my face until Jonathan rose, walked over to me, and grazed his fingers across my wet cheeks.

The same electrical current that had raced through Dierdre's hand to my body now coursed through his fingers to me (or from me to him); his eyes widened, telling me he felt it too. With the setting of the two suns, dusk cast shadows all around us and made it possible for the glow from my amulet to show on his face and in the pupils of his dark, pained eyes.

"Adrianna? It's glowing?" He asked rhetorically. We both knew that now that the secret was out and honesty was flowing like a waterfall, our connection was truly formed and strengthened. We had only touched on our potential before—now we could finally share the inner attributes that might aid us in our quest. (Yes, now it is *our* quest.) The amulet that I had never really given a second thought to now became an object of huge significance. It glowed when a connection was made: an extraordinary, vital connection like I had with Dierdre, and which I was now having with the man that I knew I could not deny and loved with my entire soul. Somehow, the amulet knew, or felt, or soaked in power from its surroundings to shed a light of wisdom.

"I can't allow this, Anna. You have shared your genuine goodness with so many who have wronged you, and I will not accept it from you after what I have done. You were right: I am a coward, and I did not tell you in the beginning because I was afraid. I wanted you here to save us, but not only that—I wanted you here to save me. To be the key to keep me from any further retribution for what I have done in the past." He saw the hurt curiosity on my face and stammered for a quick explanation.

"If I brought you here to save us all, then I would be honoured for that; I would be the one who initiated the process, I would be thanked and, well . . . redeemed. I learned a long time ago that you were unique, long before there was even a suggestion of the doom we all now face. You were known to many as the "one who would save us," but you couldn't be touched then. You were just that lonely, beautiful little girl trying to help her sister and everyone else around her. I watched you grow and felt a need to protect you. Do you remember the day your sister and you were shopping for your drunken mother, when that car came out of nowhere and almost hit you? I was there, and I somehow was able to push on the brake for the old man, who didn't even see you standing there in the parking lot."

I did remember that day. I was only 13, and Steph was so little. My mom had sent me for coffee. She had one of her terrible hangovers and would not leave the house. Even then, I was surprised the car hadn't killed us both. He had saved us. He had always been there.

"And the time you had lost your watch, from your grandma, I think, and it appeared one day on your dresser? I found it; Steph had borrowed it for one of her dolls for a tea party and it had fallen beside her bed. There are so many times, Anna, I looked in on you, watched you sleep and play and grow into who you are now. I have loved you from the first time I saw you at the park."

I couldn't help but smile at his words. To know I was loved by some stranger who saw my idiosyncrasies, my foolishness as a child, my messed-up hair, and my messed-up life. Someone who had been there for me, saved me, helped me long before I knew, and even now. How could I stay angry at this boy who said he had loved me for so long?

"Stop, Adrianna, don't smile at that, don't smile for me. I do not deserve the beauty of it, the purity you are made of. Remember, despite that love that felt so overpowering and real, I betrayed you, hurt you, and wanted to use your graciousness for my own wretched gain. I do not have the right to love you. I should have left you alone." He sat again, putting his dark head in his soft hands, shuddering and cringing at his own truth. Stunned at his maturity and destructive self-loathing, I was lost for a rebuke.

Some of the earlier hurt and fire inside flared again with the honesty he gave. I did not want to see him in that light. I could not absorb the idea that he could love me and yet use me for his own gain. Was that really how it was? Or did he perceive it this way to fuel his own fire. Suddenly I didn't care. I didn't want to know the *why*, the *how*, the *what*. I wanted to tell him that he was forgiven and that whatever reason he had for the way he had done things was good enough for me. But there was something I needed to know first, even though I would not allow his answer to sway me from my present course.

"Jonathan, do you know who killed me?"

His whole body shook like he was struck by an unseen bolt of lightning, and for a moment I really did think he was going into the canyon at his feet. I regretted the question but not the curiosity behind it—I did need to know.

"I . . . um . . . yes, Adrianna, I believe I do." His affirmation stung like a wasp intent on the kill, but there was something in his voice that questioned his own knowledge on the subject. I wanted to ask what he knew: was it magic someone had used, or did he get one of the dark angels to do it? How the dead killed the living was a mystery to me, and one I would wait for an explanation of.

"Listen, Jonathan, I know your mind does not want to hear this, but I need you to pay attention to me. I am hurt and angry at you. There were ways things could have been different, but who am I to judge and dictate to you how it should have been done? I know that you did what you needed to do based on what you felt was right. That may have been tainted with selfishness for you, but it was for the greater good, and whatever happened in your past to make you feel so strongly about deceiving your own feelings and the feelings of others must have been bad enough to warrant your actions, whatever they might have been." Deciding closeness would emphasize my words of forgiveness, I sat down beside him. He stiffened slightly at my close proximity, which hurt my feelings again, but with reluctance I understood that he felt unworthy to be near me (what the hell was up with that, anyway?).

Feeling the tension between us saddened me and I wanted to lessen the gap and tear down the wall that kept us emotionally apart. But I wasn't ready for that. Talking was the path I was on and would stay on for now.

"With that being said, I think there is a job to do and I cannot possibly do it on my own. I need you, Jonathan. Please help me?" The pleading in my voice caught his attention and for the first time since I found him here, his eyes completely connected with mine. Fear rushed through me at the incredulous look on his face, like I should not be asking him for help, like I had no right—but no, that couldn't be it, he wouldn't feel that way. He felt like I didn't need his help, that he had done all he could do and I was gifted and strong enough to take it from here. (There we go with the "Adrianna is all that" crap again).

"Adrianna, I have done enough—can't you see that? Please just leave me alone." He was defeated. There was no fight left in him. I had to find a way to re-fuel his energy, re-stock his pride, and beg him to follow through. I was terrified to go on my own; I had no idea who to talk to, what to do, or how in hell (pardon the pun) to save the world I had just learned to call home. I also didn't want to lose the man I had just learned to love.

"Jonathan, don't do this to me. Don't leave me now when I need you the most. I don't want to beg, but I will. Please Jonathan, don't abandon me." The words had more of an impact than I thought they would. Tears had formed in his eyes and to my utmost surprise were spilling over on his pale face. Shocked into silence, I leaned over to wrap my arms around his shaking body. The energy was there again but increased, almost painfully so. He seemed to feel it too, as his shaking became more pronounced as well. Touching his right cheek with my hand, I wiped away the tears there. They had slowed as the minutes passed though the trembling had not.

With a voice drained and lost, filled with an emotion I couldn't begin to comprehend, he spoke.

"Abandon you? Would I be abandoning you?" There was no anger, no sarcasm, no spirit, just pain in his question—a pleading for me to put to rest his ultimate fear, though I did not know where this fear came from. I struggled to find the words to calm him, to persuade him he would not let me down, and he would not fail or abandon anyone ever again. I wanted to ask him where this unbridled reaction came from, this terror of deserting someone or something, which I was sure was what caused his colossal inner agony.

"You will not leave me or desert me, Jonathan; we will work together, as a team. You brought me here because you believed in me, not because you were selfish. You knew I would be needed and you saw something in me I did not, that I still don't. Kindness, compassion, and friendship is what you have shown me since I came here and no matter how many spaz fits I had, you still stood by me because you are a wonderful man." I realized as the words came tumbling out like a landslide that I felt and believed each syllable. He had always been there for me and no matter what ulterior motive he may have had, he still went beyond the call of friendship to be there for me. The rush of emotion came over me again until I felt like I would drown if I did not release the dam of feelings I had working like an ocean undertow keeping me just below the surface. I stood and half turned away from his motionless form. "I have had a wonderful time each day with you, Jonathan. I can't thank you enough for all you have shown and shared with me. But honestly, I don't want to thank you, I want to believe you had to be with me almost every minute, that you couldn't stay away from me, that your resting time was consumed by thoughts of me." My own tears were dropping steadily now from my burning eyes—the dam was released, though I still felt strangled by a lack of air. I could feel him staring at me, so I turned to meet his still-moist but no longer weeping eyes. His face was still pained but it had changed; he didn't seem as lost or devastated, like I had given him hope again and a purpose, and re-set him on his chosen path.

"I have been very self-absorbed, Jonathan, and I hope you can forgive me for what that has caused. I have not thought about what being dead has done to you and where you might have come from. I have been ignorant of how hard it has been for you to keep this from me to protect my feelings, all the while knowing what the risks were in being gentle with me." He was shaking his head slowly as if astonished at my revelations. Embarrassment burned my cheeks and my tears felt warm running down my chin and neck. I had to finish. I had to release the pressure building in my chest, threatening to explode.

"I not only have been selfish, I also have been keeping things from you. I have not told you how important you have become to me—so much so that I am not sure I could imagine any life, especially a never-ending one, without you."

There was nothing left to say. Neither of us felt a need to break the silence; it was open, raw, but somehow comforting, and I could breathe unrestrained once again. Standing now, he slowly moved towards me, reaching his gentle hand to my face. His touch was warm and powerful, filling me with a sense of ease and fulfillment. I was happy when he touched me—I needed his caress. His fingers traced along my jaw line and up to my hairline, bringing my hair behind my ear then off my neck. I felt a cool wind on my warm, still-moist face. His hand tenderly travelled down my neck to the amulet he had surely been the one to give me so long ago. I knew it was glowing again; this time I thought I could feel the heat. I was learning how to be sensitive to this since the woman in the common room had first told me it was illuminated. I could again see its reflection glow in his eyes.

Amazing as it was that just some time ago I was a very ordinary teenager wanting to escape my world (not quite expecting this one), I was strangely comfortable with all the changes I had made in myself and the additions to my personality. I had always been intuitive (when I wasn't having dumb moments and lusting over college boys); I had always been friendly and caring. Now these traits seemed to have been highly intensified and were for the greater good instead of just for a few people in the living world.

I constantly felt overwhelmed, but even that was fading into an acceptance that things were meant to be this way. I was not special, really; I was what I was supposed to be to accomplish the goals set forth for me, kind of like the theory that people and animals, over time, were made a certain way to suit their environment for survival. (Thank you Charles Darwin.) But I was equipped with these instincts and tools not only for my own survival but also for the continued existence of many others and the world they inhabited, and of course in a much shorter time.

This awareness no longer scared me, although I do not know how I came to this comfort without conscious effort. However, I was certain it had much to do with seeing Jonathan struggle in his heart to be true to me, despite how much pain that truth could cause us both, and seeing his agony at having hurt me and his ability to finally let loose his inner feelings for me (even though I was still amazed he felt that way about me—yeah, just plain . . . me).

It also came from the vision of Dierdre, so small and fragile, being so strong and resilient. Her past haunted her, her handicaps restrained her, but all the while her inexhaustible patience, charismatic tranquility, and shared supernatural intuition overpowered all else to serve its purpose. Yes, its purpose. Dierdre was the key.

Shock at my realization must have shown on my face, for Jonathan's expression changed from one of gratefulness, sensitivity, and love to one of curiosity. I did not know why I was so sure, but I was: Dierdre must be the hub, the central part to all of this, and I needed to talk with her again. I also needed to learn more about the dark angels that Shadow spoke of and find them if they existed.

Jonathan looked lost as glimpses of my thoughts flashed across my face. "Adrianna, I can't keep up with you. One minute you hate me, then you can't live without me, and now you're off somewhere else. I don't know where to begin." Seeing the desperation to make things right on his face gave me the incentive to slow down my thoughts and backtrack. I didn't know how much time we had (though I assumed it wasn't much, considering Shadow's unnerving appearance), but I also knew it was imperative that I give Jonathan the support and the time to move forward with me. I was impatient. Then I looked at his face—I mean really looked. A little boy stared back at me, a timid, fragile boy, with the weight of his world on his shoulders (he was still damn gorgeous, though), gradually re-gaining the affection I once saw in his piercing blue eyes, toughening around the edges, positioning his back straight and strong.

I was in love with him. It was there all along. The voice who guided me through the dark and the saviour that never left my side.

He spoke before I could, suspecting that my thoughts had yet again shifted back to us.

"Adrianna, I have fallen in love with you completely. I always loved you but now I love you, who you are now, here, with me. I failed you, and for that I will be eternally sorry. But I can't deny it to myself anymore, I do truly love you. Your compassion, intelligence, beauty, spirit, exuberance, and your strength—most of all your strength—it all amazes me." Whoa, I was gobsmacked again, hearing the words I never thought I would hear as a dead girl at age 17, desiring those words more than I thought feasible, and reciprocating the sentiment with every part of my heart and soul.

"I do believe that I love you too," I cried with more affection than I had ever felt towards any living soul.

In an instant, we were in each other's arms, the energy flowing freely between and around us both, warming us, soothing us, shutting out the perils of the world, of our own amorous unification. There we stood, wrapped in a solid embrace, protecting each other from the pressing danger, from hurt, pain, and loneliness.

"I don't even know how it is plausible that we feel this way; I am afraid it isn't real, that it is a result of the fear of dying again, or of being in a foreign world together. What if it isn't real, Jonathan, what if it ends?" My uncertainties burst out before I had a chance to sugar-coat or rein them in.

"Anna, slow down, you're fretting again. Listen, I don't know how it is happening either. I have never felt this way about anyone, living or dead. I was informed many times that love does not happen in the middle world. Well, that it is not advisable, anyway. I knew I felt more for you than I wanted to admit the first time I saw you and then certainly after I first met you in the mortal world. I did not think I was capable of love and I was confident that no one would return that feeling for me." The incredulous look he gave me made me smile and shake my head in disbelief. HE was surprised? (I'm all that and a bag of chips again).

The sincerity on his face was overwhelming and the tears started to flow again. I didn't know what all this was going to mean to either of us. We were hurt, both scared and both probably going to die again. Now we had declared our love for one another, which we had each been repressing for selfish reasons, and there was nothing we could do about taking that love to the next level whether we died or not. What a confusing world we live and die in.

I didn't want to express my negativity to Jonathan, as he still looked very vulnerable and shaken. All I wanted to do was hold him until both of our tears ran dry and our hearts began to mend, though I had an awful feeling we didn't have that much time. Unstitching myself from him with regret, I stared again into his lost sea of blue.

"Where do we go from here, Jonathan?" I asked the inevitable because nothing else would come out of my hesitant lips. I knew he didn't have any more of a clue than I did but we needed to figure something out and fast.

"Adrianna, I don't know. I am so sorry. I will tell you everything I know from the beginning, and after that I guess we will need to talk to Shadow and Valerie too; they seem to be the ones who know the most.

Somehow, we will figure this all out—I won't bail on you again, Anna, I promise." The love in his eyes was beautiful and made me want to forget everything from my past and present. I could only see him in my short future.

"We will do this together, Jonathan, with a little assistance from our friends, I hope." I was worried that it was too late.

Dierdre. Her tiny, precious face popped into my mind like a movie screen close-up and I remembered what I had been thinking about earlier. She had to be the key. I decided to explain everything to him so that he was up to speed with the limited information that I could provide before he told me his side of the story. He looked at me quizzically, his eyes begging for an explanation.

"Ok, so I don't know how much of this you are going to understand; I don't even know how much of this I understand, but I will tell you what I know and why I think Dierdre is so important." He nodded in agreement and a wave of hope washed over him. It was comforting and gave me more certainty in my voice than I surely felt.

"The day we took Dierdre to the meadow, when we arrived back you saw us embrace and the after-effects, yes?" I knew he had but jogging his memory would help in the details.

"Well, I had just planned on hugging her but what happened was absolutely crazy. When she touched me, or I touched her, whatever . . . I felt a similar energy rush through me as I do when you and I touch." I watched his face for the reaction I was hoping for. It came: the realization that we both felt the same sensation when we touched, like two magnets being sucked together and bonded with an irresistible force. I was so pleased that it wasn't just me that felt it.

I explained the vision that Dierdre had shown me, watching the expressions on his face, which mimicked my emotions.

"It was awful, Jonathan. I don't know what happened exactly or what that sweet little girl has been through, but I sure know it wasn't good."

I let the scene settle in his mind before I continued. His features were hard, lips drawn tight, body rigid. He felt the same as I did. How could anyone hurt an innocent child?

"Dierdre was just as shocked as I was that the vision happened like that and she tried to apologize to me in sign language. The man who I later learned was Peter told me what she was trying to say. Afterwards I went to rest . . . well ok, I was supposed to rest, but I got anxious. I got up and started pacing, thinking about her, and wondering how I could possibly help her and learn to communicate with her. That's when I saw the books and everything else." I laughed at my memory of finding the exact literature that I thought I would need to learn about her. Jonathan smiled at my undisclosed joke.

"I had a new bookshelf stocked with books, and a few other new additions to my room. So, I settled in to read for what seemed like forever. Then I gave up and went to see Dierdre." Curiosity was burning in his eyes, darkening them to a deep gray-blue. I would never get my entire story out if I kept looking at him. (He was heavenly.) I went on, looking more at the deep canyon we were teetering on the edge of than at him, and told him every detail about my connection with Dierdre, our new-found communication, my amulet glowing and my realization of her uniqueness. He nodded, smiled, frowned, and looked puzzled throughout my tale, but in the end, I knew he agreed with my perception of the amazing little girl.

"She can talk, Jonathan, but she chooses not to and I want to know why. Somehow, I think it is vital to know what happened in that home and to her in her past life, and why she is bound here with us. Why has she been chosen to stay behind, to join with us to save a world she is not even supposed to be in?"

"I agree with you, Anna, it doesn't make sense. No one has really ever questioned it—she was just here, a pleasant addition to our atmosphere." Cringing at his own words and the seeming ignorance behind them, he quickly grabbed my hand like he was warding off the slap he thought he deserved for taking her for granted, as they all had.

"I'm sorry, I don't feel that way about her, and I guess I never really did. I always wanted to be her friend and I spoke to her often. She always seemed to like me." The wistfulness remained on his face through his silent memory of their meetings and moments together. It was plain to see that he adored Dierdre then and would do anything to help her.

As he was lost in his own thoughts of Dierdre, I became lost in my own, suddenly remembering the pictures she had drawn. The one of Jonathan and me together was comforting and a little invigorating. It gave me hope that our intimacy would progress and our union would prevail.

The one of me as an angel was confusing and left me slightly unsettled. I did not see myself that way and was quite uncomfortable for her or anyone else to either. It gave the impression that I was all pure, holy, and powerful, and that I would conquer evil and become victorious for all souls.

However, neither picture stirred the kind of emotion I now felt again, envisioning the last picture—the picture I wanted to pretend didn't exist. How could I possibly be walking off, back straight, hair flowing, shoulders squared with confident acquiescence, while Jonathan sat, weeping painful tears of loss and rejection? I couldn't, I wouldn't do that to him. I would not leave him now, especially after catching such breathtaking glimpses of the true man behind the strong, untouchable facade—a lovingly intrusive glance into his heart, where I found myself, Dierdre, and many others that he had yet to reveal to me. He was complex, sentimental, and more intriguing than I had imagined he would be.

So, NO: there was no way that drawing had any merit. She must be confused about that part of our future. I decided to keep that part from Jonathan, though I told him that I had seen some of her drawings. He seemed overjoyed that I had.

"I saw some of her creations before as well. They are utterly amazing. I have seen art but that is incredible for a child that age. She truly is gifted and it's . . .." He trailed off, not wanting to finish the sentence with the words we both knew were fitting. It was horrible to think of Dierdre's being dead and bound to nothing for eternity; if there was to be any proof that reincarnation did take place, I would gladly accept it if it meant that she could move on and become a great artist in her new life, a life that did not begin in pain and fear.

"Well, you must be right: there is more for her somewhere, and we just need to find out what to do. But I still want to talk to Valerie first. Why don't we head back?" He took my hand, and we both walked along the flat, arid land that perfectly depicted a scene from one of my other favourite movies, *Star Wars* (I always loved the name Anakin, and was set on naming my first son after him). I could almost hear the motorized *whoosh* noise of the speeders and the rumble of the Jawa land-crawler. (Yup, I was losing it... marbles falling out of my head.) I gripped his hand tighter as we walked, hoping to squeeze out some of the insanity creeping in. We both needed a mental rest and neither of us was going to get it. Heaven only knew how long it would be before we could casually bring up our open love or our irrational hopes for a future together.

Jonathan must have decided along the way that the regular trek on foot would take too long, because we were suddenly standing outside the common room door listening to a very loud wail coming from within. This couldn't be good. Glancing at each other simultaneously, wearing identical expressions of apprehension, we entered together to find chaos at its worst.

Many were crying, some sat or stood reserved and motionless, all wore a look of devastation and impending doom. Something terrible had transpired, that much was sure. Intuition screamed at me that much had changed since we had left and Dierdre might need us more than we knew. I searched desperately for her, trying not to make eye contact with anyone I scanned over. Avoiding the pain and uncertainty in their eyes was difficult, but I knew finding her was crucial.

Dropping Jonathan's hand, I darted for her little table to see if she might be hiding under or behind it. She wasn't there, but what was there chilled me to the core. Another beautifully crafted addition to Dierdre's creative portfolio was lying on the desk. Emotions shrieked from the page as the scattered forms became a complete vision there. There were figures with agonized expressions, mouths open to yell out unheard terror. They were faded like Shadow. There were rough parallel lines branching thickly and sporadically from the ghost-like people in an upward direction like they were being sucked towards an implied sky or abyss. Other forms sat obliviously on the floor of another room separated by one crude wall. They were spaced and positioned erratically, like there were many more of them there, and at least half had been chosen, randomly, to be erased.

Fear blazed through me, leaving me feeling charred and raw to the elements. Before I even saw him coming Jonathan was at my side holding an arm around my waist, as if he knew that I was seconds from collapsing into Dierdre's vacant little chair. We both knew instantly that this picture was not a premonition anymore—it was the close past. It had just taken place in our world. The atmosphere in the room now became clear, as if a movie narrator was telling us the events that had just unfolded in our absence.

I knew without a doubt that Shadow must be lost to us forever and that others had left at the same time—most likely the elders, being the longest inhabitants here. I also knew that the other room in the picture was the suicide room, or the lost room, as I had learned to call it. What attributes had chosen them, I didn't know, but many were gone without the awareness of the souls remaining.

Kieran, my brother, was in that room, although I was not allowed to visit him in there and I knew it would do me no good even if I could. I had taken a small, though disturbing, comfort from the fact he was here, in the same world as me, and thought that if I could figure out a way to save the lost souls, I might just save him too. But now that chance seemed a distant dream gone wrong. I could not save him or anyone else anymore.

My body drooped in his strong arms, hitting the chair behind me gently. We were too late. People were leaving, evaporating, permanently ceasing to exist, and not by choice. Wondering if anyone had been informed of the cause of their doom, even the weak explanation of our demise that only a few of us had known, I searched the room for expressions of confusion and disbelief, but all I saw was misery, defeat, and sheer panic.

At that moment, a few of the occupants became aware of our presence as if we had jingling bells around our necks. Apprehension filled me as the same lady who had noticed my glowing amulet approached us. She had been crying and looked as if it was her first day here and she had just grasped the fact she had died.

"You haven't been told, have you? You don't know what has happened here. We are dying again. The elders are all gone and Frank, the salesman from Kentucky, he is gone; Lori-Anne, that sweet woman who was creamed by a transport truck just a short time ago, she is gone too." Her voice was shrill, fast, and relentless. She was not in control of herself and I wondered if I should tell her to stop speaking and calm down. (It wasn't like she was going to have a heart attack or anything.)

"There are others, too. I don't remember their names, and they say many from the suicide room have gone too. We are being cancelled out and there is nothing we can do but try to move forward. I don't know how. I can't seem to do it." She sobbed loudly until almost everyone in the crowded room ceased their own babble and tears to listen silently to her gut-wrenching cries. I knew we had to do something, and fast, or there would be nothing left to save even if we could find out how to proceed.

I turned to look at Jonathan and realized suddenly that he wasn't beside me. My body trembled in fear. I couldn't bear this alone. I wanted to know if Kieran was gone and where to find Dierdre (not allowing myself to consider that she could be gone too) and Valerie; we needed to speak to her as well. Just as my panic threatened to take over and leave me wailing like the unnamed lady, I heard the voice I needed more than anything else, the voice of reason of my past, and now my future.

Jonathan had stood up atop one of the long bench tables up against the far wall so that he could look out and down at everyone there. His voice was loud, direct, but full of compassion and empathy. He quieted the room at once.

"I know that everyone here is scared. We have lost some of our world today, and you all have been told that we will all depart this life like the others that have been taken. You are afraid to stay and face an eternal death but have no knowledge of how to "move forward" into the existence you have only heard about, the one where everyone lives forever without contact with or the ability to see the living world, the mortal world. This place to move on to has been talked about, as has the opportunity to reincarnate and join the living world once again in a new form with a new life." I wasn't exactly sure how Jonathan knew what everyone was thinking or where he was going with his charismatic words of understanding and wisdom, but he sure had the attention of every soul in the room and more had joined us since he began.

Grinning with all my might I nodded to Jonathan to follow the direction of my outstretched arm—I was elated to see that both Valerie and Dierdre had joined the congregation just moments after the speech had begun. Valerie's frown and look of abhorrence could not dampen my joy at seeing for myself that they were ok, no matter how sad they must be. He did not pause long, but long enough to briefly nod to let me know he saw them in the back. I decided to make my way over there, not to talk but to just be near Dierdre until Jonathan was finished with the room full of terror and sadness.

"I cannot tell you if that is true, if we can all move on and reawaken somewhere else as someone else. And no one else can tell you that either because no one knows. But I can tell you that each one of you has chosen to stay here this long, for whatever reason you may have had: you have friends here; you want to be close to your loved ones on earth or to simply watch over the world as an angel in the sky. Whatever your reason, you chose to stay here. Now I want you to make that same choice again." Gasps rang out in the room from all directions and a constant indefinable drone began. I again wondered what Jonathan was hoping to achieve with his methodical words, but I had faith that he would have his way and all ears would be his once again. He seemed to know this too because his confident gaze into the crowd did not waver.

"I do not ask any more than you have already given. But I have a reason for asking this of you now. I believe that whoever told you of our impending demise was wrong." Louder gasps this time and rude words of craziness and ignorance came from select places in the audience. Jonathan stood unwavering.

"For quite some time now I and a few others have known that our world might face a time of peril, though we did not know in what way it would come. We only knew this when the introduction to a very special person came about by accident. We knew that this person and a few others were to be the ones to help us all in our time of challenge and need. Our world was not created, so many thousands of years ago, to be abolished without our having a way to fight against it or prevent it all together. We do have some tools to fight this destruction, and I beg of you to help us fight and stay together as we always have been." The murmurs were loud but now filled with curiosity, and some with hope.

It was the consensus that they all wanted to stay here as they had wanted to do since they arrived. The connection, however minimal it was for most, was powerful enough to stop many from pursuing the option of a possible re-birth. But that was relative to the fact that in the middle world there were believed to be no more than two hundred people. More than that died on earth each day and they obviously went somewhere other than here. Who knew how many realities or dimensions there were? Maybe the day that some died was the very day they were reborn as someone else. It was all so huge, complex, and bottomless. (I am so glad I am not a theologian).

"I cannot force you to stay, nor can I tell you who the special ones are or what has to be done. I just ask you to make the choice once again and leave the rest up to destiny. We will mourn the loss of those who are gone but then summon up our courage to continue as we were, knowing that standing together will help us all stay here together." When he finished speaking, questions were being thrown at him like balls from a batter-up ball machine. He dodged each one curtly and made his way to where I stood, close to where Dierdre and Valerie had been a moment before he reached me.

Taking his face in my hands, I kissed his soft lips, hearing the surprised noises from the closest bystanders.

"That was great, Jonathan. You gave everyone understanding, hope, and freedom, which is a whole lot more than they had when we first walked in. Now why do you suppose Valerie was so disgruntled and took off with Dierdre again?"

"I don't know, Anna, but I intend to find out." Grabbing my hand, he spun me towards the nearest exit. Just as we left the room, I heard the exact words I did not want to hear but had expected to anyway.

"I bet that Adrianna girl, the one with the glowing charm, is the special one."

"Yes, you're right—that must be why she looks like an angel."

The last comment threw me for a loop and I blushed in embarrassment. Jonathan had also caught the fleeting words and noticed my reaction.

"I am not the only one who believes that." His words touched me with their sincerity. There was no way I could deny my love, respect, and adoration for this boy. It was like a constant, wonderful dawning each and every time I felt the unusual emotion come over me, like a feather-light squall swallowing up a dry-docked ship. I absorbed him, his every word, his every touch, and what an empowering feeling. (Wow am I sappy.)

We walked on, looking for any sign of where Valerie and Dierdre might have gone. I couldn't understand why Valerie was suddenly attached to the child she had never really held any interest in before. I had to admit to myself that I was scared for Dierdre. Not that I didn't like Valerie, because I did, but I couldn't shake the feeling that something was amiss with her toting around Dierdre.

Just as I was about to voice my opinion to Jonathan, around the corner came Juniper, looking even more dishevelled than the rest of us. She looked wasted away—not faded as Shadow had been, just worn thin like she had lost her own will to continue. There was an almost wild look in her eyes, in contrast to her slow, fatigued body language. She glared at Jonathan and me like we were the last two people she wanted to see here. It dawned on me then that her husband, the one she had loved and lost so quickly, had been in the same impenetrable room as Kieran. I was filled with dread that quite possibly he was gone. The words were out of my mouth before I could get a hold of them.

"Oh Juniper, I am so sorry. Your husband?" Jonathan glanced at me, shocked at my unusual abruptness.

I thought for a moment that Juniper was going to attack me. Her large-for-a-girl hands were balled into fists and her already-pale face turned snow white. I immediately regretted opening my mouth and jumped to apologize. She beat me to the push.

"No, you're wrong, sweet Adrianna. He is just fine. Still in that godforsaken room with all the other lost souls. No, he is just fine." Her gentle words contradicted the strangely deformed mouth they came from; I wasn't even sure how she could form her words so carefully with the right amount of niceness through such a grimace. Well, if she hadn't lost her husband, she must have lost someone dear to her to be so dishevelled; or, quite possibly, it was her own transience that concerned her. Either which way, she was upset.

"Oh, well . . . I am . . . glad that . . . well, crap." I stumbled to find the words to express that I was happy he hadn't been taken from that awful room full of lost souls who didn't have a chance at knowing anything but their own mental oblivion, but that sounded stupid. Would it be better for them if they did just disappear? I certainly didn't know, but I did know I would want whatever was most comfortable for Kieran.

"Don't worry about it. I understand what you're trying to say. You feel the same way as I do, with your brother being in there too. But they are both fine. I made sure. Now I am off to do some damage control, if there is anything left after that heart-warming speech you made in there, Jonathan; if only it were all true." Her laugh that followed was bitter and did not match the kindness she was trying to portray. I could see Jonathan preparing himself for an argument, as he did believe that all he had said was true and that the effect it had had would be genuinely felt. I was still a little dazed at her knowing about Kieran as she waved abruptly and started walking down the hall.

Jonathan's fight fell silent, the rebuttal dying on his lips. He looked perplexed at her attitude but happy to forget the altercation altogether; I, however, was still a bit tweaked. "Wow, she was freaked out, don't you think?" I asked, a bit baffled at her demeanour.

"That's just Juniper. She has never been the calmest, most rational sort. Don't mind her, let's find Val and Dierdre," he answered quickly.

"Yes, that's what I was going to say before: isn't it a little odd for Valerie to be taking such an interest in Dierdre?" I hoped my question wouldn't offend him, as I was sure that he thought a lot of Valerie. She seemed to be the one he went to with questions and for advice. Although she was quite crass and forthright, I did like her. Maybe all the stress was making me overly sensitive and paranoid. I wasn't usually quite so judgemental— ready to jump on Juniper's actions, and now fretting over a psychic hanging out with the only child here, who had obviously gone through a substantial ordeal today (it appeared her whole life was an ordeal—and I thought mine was bad).

Relaxing my shoulders and thoughts, I sped up to keep pace with Jonathan, who seemed to be racing. I wondered if I had shot some of my suspicion his way. That made me feel bad. I intended to apologize but he jumped ahead of me, shouting back.

"It is odd, Anna; I don't like it one bit. What could she possibly want with that poor child now after ignoring her for so long? You know, I tried before to get Valerie to talk to Dierdre, to befriend her in hopes we could learn more about her past and her death. No one knew how she had come to be here, let alone come to stay. So, I thought that her being a "seer" might make her the perfect candidate. But she refused—stoutly refused. She didn't want to touch her, or try to see anything about her past or future. I don't like it, Anna. We need to find them." He was already way ahead by the time he had finished speaking. I couldn't imagine why anyone would hurt a sweet, innocent child, and if there were such a person, I certainly wouldn't imagine it to be Valerie. I intended to find out just what was going on.

We travelled an unknown distance, searching, exploring each room we came to. There were many rooms I had never been in. We trudged on. Despair started to fill me like an over-flowing cup of hot tea. Dierdre and Valerie could be anywhere, and seeing that I had no perception of how large this world could be, I shuddered at the thought of the time it might take to find them. Jonathan was tense as well. His frown deepened to the point of anger after each empty room. Eventually he slowed to match my stride again. I was grateful for the speed change and wished we could rest.

"I think we might have to go back to the common room and ask if anyone has seen them. I also want to check in on what is happening and how everyone is coping." His face once again held a look of defeat that bit into my soul. I had not until this moment stopped to think of how the loss had affected him. He had lost his friend Shadow, and although I did not know Shadow well, I did know that Jonathan had looked up to the old man for advice, common sense, and tactful criticism. I also had no idea who else had been lost and what they could have meant to him. Tears welled in my eyes at the thought.

"Adrianna, what is it?" He rushed over to my side, taking my trembling hand in his own warm, safe grip. Worry clouded his eyes and some of my sadness was replaced once again by disbelief in our feelings towards each other.

I stuttered, trying to organize my rushing thoughts. "I . . . ummm . . . well, I am sorry, Jonathan. We have been so busy looking for them, and with all the chaos I didn't think of how sad you must be to have lost your friends today. I can imagine how you feel—well, I guess I know how you feel." The pain of losing my own friends and family was still brutally fresh in my heart and mind, a pain I was sure would diminish in time but never leave me completely. Then there was Willie . . . .

"I am sad. I will miss Shadow; he was like a mentor to me since I came here, and one of the only ones who calmly tolerated my arrogance and outbursts. He was also one of the first to realize that I could do things . . . things that others couldn't. I really was quite a pain in the ass, Anna, to many here. Looking back, I wish I hadn't been but I can't change that now. I only hope that Shadow knew I meant no harm."

Wistfully he stared off to a memory I could not see. I hoped it was a pleasant one that would help him to remember a happy time with his aged friend.

"I am sure he did, Jonathan; he knew you well enough to know . . . well, to know how good you are inside despite your sometimes-awkward way of expressing yourself." I hoped I hadn't slighted him, but quickly realized I had not when the familiar, humorous sparkle appeared dreamily in his glamorous blue eyes.

"I wish I had known him just a little more, but what I did know proved him to be a very wise, very kind, very insightful man who honestly wanted the best for all. A man full of honour." I said.

"On a slightly different note, what exactly are all your powers, Jonathan?" I asked, half curious, half wanting to change the subject to a more uplifting and confidence-building one. I was sure I knew most of what he could do, especially the things that I could not.

He smiled softly, having known this question was coming. He seemed prepared to answer even at this inopportune time, in the middle of lost comrades and missing friends. He answered my question anyway, with a delight that soothed and intrigued me.

"The elders believe that some of us who arrive unknowing and unwillingly here in this world bring something with them. A personality trait, a talent, an extension of, or the opposite of a way of being, shall we say. Not all of us do, or some don't recognize what followed them through, but like you I brought a few things with me." I heard his words but I was having trouble understanding exactly what he meant. I did not see how this could relate to him before and now, or to me for that matter (I was a very plain-Jane, ordinary chick before).

He recognized the look of confusion on my face and continued with examples to help my understanding.

"Valerie was very intuitive as a mortal; she lived as a fortune-teller during a time when some "seers" were murdered for their visions and taken to mental institutions for their talents. She obviously brought that with her. The old man who knew sign language was given back the gift of hearing and speech, and Shadow brought the knowledge of history, the Bible, religions, and the ability to have serenity in times of peril. There are many of us here who are special in some way." He appeared to be remembering something of great importance to him, as his eyes glazed over and he gripped my hand tighter.

"I had many . . . let's say not-so-sought-after talents before I died. In some ways, I was plainly arrogant, pushy, selfish, determined, and lost. I had so many ideas, so much energy and passion, and nothing to do with these. So, I bent and broke rules, upset many people, and let down the ones

I cared the most about. My father was not often home: he left the store to work away, trying to support my mother, me, and my siblings. My two younger siblings—a sister and a brother—and my older brother were quite different from me. They were happy in the little town we lived in and the small house that needed repairs, having little to no money and living and working each day for the same small goals: food on the table and warmth in our home. I wasn't happy with that. I wanted more." He paused for a moment to swallow and clear his throat. I could tell this was hard for him. I could almost feel the tension rolling off him.

"My father stopped coming home after a while, so my brother went away to college to learn carpentry. He wanted to come back and fix up our home for my mother and us; that was his plan. It broke my mom's heart when he left, then left a hole in her heart when he was killed in a construction-site accident. There I was, the only male figure around to do all the chores and care for my mother, sister, and baby brother. I was fourteen at the time and spent most of my time raking leaves, shovelling snow, running errands, and digging holes for fences for what little I could earn. I resented the position but loved my mother. She was a wonderful, beautiful, caring woman, you know—a lot like you are, Adrianna. I loved her and wanted the best for her, despite my own dreams and aspirations."

A pain I didn't understand had plagued his voice and saddened his face. He was speaking to me but he was not with me anymore. Somewhere in his past, the little boy with the bright blue eyes was struggling to find his way through the absence of a much-needed father, the loss of an idolized brother, a life of poverty, with shame and hardship stifling his own wants and needs as he cared for the needs of his family. Seeing this boy brought tears to my eyes. I hurt for him then and now.

"I worked, tended the youngsters, and quit school. My mother wasn't happy with this choice but she knew it had to be done. I daydreamed all the time of a different life, one with money, luxuries, new dresses for my sister, a washing machine and dryer for my mother, and wished for my father to return. He never did."

I watched him speak aloud the tale of his sorrow-filled life and regretted any complaint I had ever cried about in my own. He was stronger than I had given him credit for. His heart held such pain. I reached out to wipe away the silent tears than ran lifelessly down his hard, weary face, wanting to give whatever comfort I could. He did not shy away from my caress and I was grateful for that. Miles away, years in the past he stayed, reminiscing about a world I could not fathom and was secretly thankful for not having to ever feel. He continued slowly, with emotion-drenched words of his time, his past.

"When I turned 15 I began to get into a bit of trouble and was caught for stealing and smoking. The boys and I started nipping some brew from one of their father's stashes. We were caught and my mother was angry and sad. I wanted to tell her I was sorry and that I wouldn't do it again, but even then, I had the foresight to know that it wouldn't be the last time. I was heading down a path that I should have stayed far away from. There was a girl I really liked; she was pretty, smart, and her parents were rich. She wanted to be a doctor." He blushed slightly at the memory and met my eyes for the first time in what seemed like an eternity. His eyes were red and held the remains of un-fallen tears, though soft and welcoming. I was happy that he had some joy in his world back then. (Slightly jealous on the inside—it was some rich girl.) He must have caught a hint of the green-eyed monster in my expression.

"Hey now, she was pretty but couldn't hold a candle to you, Adrianna. She was blonde and I have always preferred brunettes and brown eyes. Yours are . . . well, the most intense I've ever had the pleasure of being drawn into. And you are far more beautiful than anyone I could ever deserve to lay my eyes upon. You are . . . ." He allowed his sentence to trail off, leaving me silently pleading for him to continue and never stop. Usually I was uncomfortable with compliments and immediately denied the possibility of their truth, but when Jonathan spoke words of such intimate description of his perception of me (no matter how unbelievable and far-fetched), I craved more and yearned to be everything he said I was.

"I'm sorry, Anna; I have been dominating the stage here with my feel-sorry-for-me speech. Please forgive me."

"No, no, Jonathan, do not apologize. I want to hear everything" (whoa, slow down there, girl). I sounded desperate and intrusive. Again, hoping I hadn't affronted him, I relaxed my pose, gazing down at my feet. (I never seemed to know when to shut up.)

We had stopped long before by a cast-iron-backed, wooden-seated bench. It was uncomfortable, but better than traipsing the same white, mundane halls that surrounded us. I couldn't figure out why no one had thought to add a little color, adornment, or a painting or two to the walls. It was so dull and unappealing. Jonathan was leaned over on his arms, his head bowed and tense. I wanted to wrap my arms around his strong, muscular shoulders and never let go. Thinking he needed a portion of his three feet of personal space, I stayed on my own side of the bench and crossed my legs, leaning back against the hard backrest.

"Thank you, Anna, for listening to me, but there is no happy ending. I will finish telling you someday just why you should not be so quick to help undeserving people like me and the others you have worried so needlessly for. But for now, I think we have spent enough time allowing me to bellyache and seek sympathy over my sordid past. Let's look for Dierdre again."

The look of self-loathing re-appeared on his face, aging him beyond his years and saddening my heart once again. The life he depicted was lonely, complicated, and heart rending, not one where he should feel hatred for himself or anyone. It was just the hand he and his family were dealt. Confusion frustrated me, adding to my unasked questions about this boy I loved.

We walked in silence for a short time, and then heard a much-desired voice coming from an open doorway to our left.

"Dierdre, you plop yourself on that chair in the corner and rest a bit. There are books in the bin, and a drawing tablet and pencils on the far table. I will return shortly." The voice was kind, surprisingly gentle, comforting, and the voice of Valerie. We both looked at each other at the simultaneous realization and quickened our pace to the entryway. We met Valerie as she was leaving, a look of hurry on her face. She stopped short just before colliding into me, Jonathan placing his hand across my chest to break the near impact.

"Jonathan, Adrianna, fancy meeting you two here. I would have thought you would be giving out autographs in the common room." She glared, full of spite, at Jonathan.

"What is your problem, Valerie? You know everyone was running rampant, and only complete chaos could have come from everyone losing their minds over what happened. I had to say something to calm them.

You might have thought to do the same, seeing that you are one of the ones here the longest, now that all the elders seem to be gone."

There was almost an equal amount of spite in his words to her. Surprised at their mutual animosity, I kept my mouth shut for fear I would say something stupid (I had eaten my foot more than enough). I noticed that she was dressed all in black, with beads around her neck and her hair drawn tightly back into a mesh bun—ironically dressed for a funeral. I instantly wondered if they had funerals here. Had anyone ever re-died before? I made a mental note to ask Jonathan later. I was never at a loss for questions, and I hoped he would never tire of my insatiable inquisitiveness.

"I did speak to everyone, Jonathan. I told them that the elders were gone, some of the lost souls were gone, and that we would all be gone before long. I have seen the end, my boy, and neither of you can stop it now. Some are to remain."

"You are the one who caused the upheaval in the common room?" He stared at her smug face with an incredulous look of his own. She offered no answer.

"Have you gone mad, Valerie? You know there is a way to stop this. You saw it, we believed it, and even Shadow knew we were right that with Adrianna we would find a way." She winced visibly at the mention of Shadow's name, the pain of loss evident on her face. I couldn't help but feel for her despite her present rudeness. I had the distinct feeling that there was a lot more going on than we were aware of. Valerie could no doubt be cruel and tactless but internally she was good and kind. What was happening here must be a product of something else. I knew it was imperative that we get to the root of this before we lost all hope.

"Informing the others will only cause hysteria. None of us really knows for sure if the ones that leave here move forward to reincarnate, Valerie. There are no absolutes anywhere. We must stick together here and work to save this world."

"Nothing is as it seems, Jonathan, and neither are you." She turned abruptly and all but ran down the hall. We both stared silently after her, not comprehending her final words. Realizing that Jonathan was trembling lightly, I reached for his hand. It was cold—as cold as the look in his piercing eyes. I shuddered involuntarily, plagued with fear and sadness. I wanted to help him and to understand why this was all happening, and wished I was the one we all so needed me to be.

Jonathan sensed my own desolation and tightened his grip on my hand, the coldness leaving his eyes.

"Adrianna, we will find a way. We must."

Attempting to console me without confidence pained him further, and his shoulders drooped in defeat. There were no words to say. Not now. Not yet. Until we had more to go on, we would be leading a blind quest. Silently, I worried that Valerie was indeed correct. All was lost.

* * * * *

Standing in the hall, the sullen pair stared into their own thoughts, into their own lost and fearful souls, defeat in their bones and on their faces. Jonathan suddenly looked up and into Adrianna's tear-filled eyes.

"I know a place; we may find some answers but we must go without being seen. I am not even sure I remember the way." He grabbed her hand and they took off to a wing where Anna had not been before. Here the atmosphere changed drastically. The walls were now made of lightly stained wood, and the cathedral ceilings made of the same, with huge, thick beams and a few stained-glass windows giving the feel of a church, an ancient church. The floor remained uncovered but was now a different kind of concrete, black like ash and softer under the step. Adrianna stared wide-eyed at the beautiful but slightly chilling ambience.

Jonathan slowed his speed, looking at each hardwood door they passed. The hall was much wider here and fewer doors appeared. They eventually entered a large room full of wooden pews behind a stunningly huge, carved altar. Candles covered the dark, old tables at the front, but none were lit. This was a church. Though there were many fewer furnishings than Anna had usually seen in any church she had entered, the pews, altar, and surroundings were the same. The air was slightly musty and cooler than anywhere else in the middle world. Anna immediately felt wrong, though Jonathan appeared to be completely at ease.

"I have only been here a few times, a privilege that most others were not given. The elders met here frequently. I think this church was a replica of the church that Shadow used to live in, although I could be wrong. There is no organ, or any Bibles, books, or offering plates. I don't know much about churches but it never seemed right here for some reason. Shadow had a private office off to the right there." He pointed to a dark-stained door in the far-right corner of the much-too-small-to-be-a-real church room.

The door opened with a creak into a very tiny, cluttered room. A large, heavy-looking cross on the far wall was as bare as the church, though the carved detail in the cross was breathtaking. In front of the cross was a small, mahogany desk bearing several large melted, lit candles, a leather-bound book, and nothing more. The walls were all bookshelves, filled with large volumes that looked as old as Shadow himself had been, and papers were strewn all over the shelves amongst candles, old artefacts, chalices, and gold-plated dishes. No pictures were hung, no windows were hidden behind the shelves, and no other furniture occupied the room.

Behind the desk sat a small wooden chair. Jonathan now took this seat and began rummaging through more papers in the three tiny drawers. Holding up a large cloth book, he opened the pages carefully. A look of fondness entered his eyes as he sifted through the beginning pages.

"Shadow read some of these pages with me a few days after I began to realize I had some talents that the others didn't possess. It is a book about the Archangels. Honestly, I was a bit bored with the tales and don't remember much of what he said but I enjoyed the time with the old man. He really was very brilliant, whether he was a murderer in his old life. He told me on that day that I was like an angel, with a quest ahead of me that would protect the mortal world. Angels are messengers and protectors, you know."

Although he was speaking aloud, facing Adrianna, his words were almost too quiet to hear and directed internally as well. The memory pained and pleased him equally. He continued flipping through the brittle old pages until a large piece of folded paper fell onto the desk. Reaching for the paper quickly, he looked into Anna's eyes to see the shared curiosity. Unfolding the small parchment, he read it aloud, expressing shock and delight at the note addressed to him.

"Jonathan, I pray that you have found this in time for it to give some advantage in the frightening situation you are enduring. If not, then let my words fill you with the confidence needed for your next life. I am gone and not overly distressed over that fact, as I lived a long mortal, and even longer immortal life. I gained much wisdom in my trials, tribulations, loves and friendships, mistakes, and victories, as I pray you will as well. I was possibly harsher with you than I needed to be; perhaps, having more time I would have been kinder and explained my reasoning's. However, time is short now, as is my patience for writing" He continued.

"Preferably you have beautiful Adrianna with you, although neither of you yet knows how absolutely special she is. Time will answer many questions you both have, as I cannot. Assuming there is a world left to save, I beg of you to make haste. Read the volumes I have on my disorganized shelves about 'The Beginning', a rather interesting tale of the first Archangels and the emergence of their nemesis Lucifer. Reading the lines, and between them, will shed much light on the darkness surrounding you all. There is one among you, however innocent in appearance, who aids the dark angels' vengeance, though I cannot give you a name. You must find this enemy, but be careful. There is also one who is the connection to this evil, though it is not known who it is.

Those of you who can "see and speak" differently than the rest of us have the answers deep inside. The visions may not come from one who is expected to give them, and some images are based on tainted beliefs. A grave, complex puzzle, to be sure. Had I more time, I would have revelled in the deciphering. Since meeting you, Jonathan, I have acquired many recent publications of this age and find many to be startlingly creative".

Returning to the peril at hand, I must beg of you to tread lightly with the hearts and souls you encounter: some are fragile, many needy, a few are broken, one beyond repair. There is sacrifice to be made, but it cannot be avoided or requested.

When I was alive and much, much younger, I cared for my church, its beliefs, customs, and rites, and in only that did I breathe, love, and truly live. I lost much along the way. Did I make the right decision? For me I did. But for others, what they truly love, desire, and cherish the most is not always evident and is easily taken for granted. Loss is inevitable if the heart is not open and the mind not clear. Selfishness is the road to loneliness and pain. I don't expect that my words are clear, but in time they will be and can be used as steps to the pearly gates (metaphorically, of course).

Adrianna has much to learn but her heart is pure. She must continue with strength, hope, and compassion, as she always has. The friendships she has formed thus far are of greater importance than she knows. I cannot stress enough how important she will become. The ones who remain here are vital, as their souls, combined, will give Adrianna more power. You must unite everyone.

Jonathan, you are a determined young man who has more latent power than he knows, but it is in a much different form than Adrianna's. I have seen within you an opportunity for immortality, and qualities that are lost to most good men and unknown to bad men. However, you are closed, stubborn, and hardened. Lessons that were not ingrained in you will be essential. There will come a time very soon when you will change the way you see yourself and everyone around you. I pray it is not too late.

Love will lead the way.

Good luck to you both. Although I am gone, I am not lost. I will find a way to be of help to both of you again in your hearts.

I pray for you both; I pray for you all.

Edward Grim (Shadow)."

Jonathan finished the letter with tears in his eyes and heaviness in his heart. Pain, fondness, loss, fear, and frustration were but some of the emotions he felt reading the cryptic note. He was certain that, when given the opportunity to read it over and over, much of the mystery surrounding his honourable words would be elucidated, but now the loss of his friend took precedence over clarifications. Mulling over the words flooding his mind, he gazed into Anna's tortured eyes. She was on the verge of tears again, still raw from the endless time of the poignant chaos that encircled her life and death; her precious face almost crumbled in a pain he could not quite comprehend. Reaching out, he took her trembling body into his arms as her composure broke and emotion flowed out in a tidal wave.

"Why am I so god-dammed special, Jonathan? I don't want to be. I don't feel special. I am a regular teenager; I know nothing of how to save anyone, nothing of angels and devils, life after death and all that. I just wanted to be me, not some saint who can't possibly live up to expectations this huge. Jonathan, I don't know how to do this and I don't know how to help you." Her agony spilled like water from an opened dam. Shaking in his arms, she felt fragile, cold, and soft. Smoothing her flowing hair under his hands, he calmed her tremors but unleashed another series of gut-wrenching sobs.

There was more to her pain—he had suspected for some time that she was still torn between her old life and this chaotic world she was flung into. She deserved a normal life, one in the mortal world where she was finding her niche. Guilt stabbed into Jonathan as if shot from a gun at close range. Almost crumpling over himself, he fought to regain his own poise for Anna.

Knowing it was imperative to hold his own and to be strong for her gave him strength from an unknown source. Feeling the power of his unbridled affection flow into his limbs and deep into his soul, he tenderly drew up her chin until her deep brown eyes met his.

"Adrianna, my love, my sweet, beautiful love. You must know you are special to me, to your past loved ones, and to all who remain in this place. Denying your true self will only hurt you. The only expectation anyone has of you is to be yourself. You have always conquered each enemy you have battled, each obstacle you have faced, and come out on top, your heart pure and unscathed. I don't know what the answers are but I know I love you and together we will see this through to the end, whatever end that may be.

Shadow, or I guess Edward Grim—I never did know his true name— has given us answers here the best he could, I truly believe that." He paused momentarily to let his convincing tone break through the shell of doubt that shrouded Adrianna. Her eyes softened before him, energy slowing, regaining composure, soaking up his confidence. Knowing his gentle words were having a calming effect on her, he continued.

"He believes in us, Anna, and so do I. We may not be victorious, we may not save everyone, but we will try and we will believe in ourselves that we are special, that we have gifts to offer that will prove our significance. Just believe, Adrianna, and we will be all right." Her forceful hug almost knocked them both into the desk. It was full of passion, acceptance, and appreciation, filling him with the same. At that moment, feeling her energy encompass him, he knew his words were true, as were Shadow's.

Organizing his thoughts, he knew he must lead for the time being for Adrianna's confidence to grow, then she with her latent leadership qualities would take over and show them both the way. He was surprisingly comfortable with this and held no resentment for her dominance, just awe and respect.

"Anna, throughout your life, what have you always done when faced with a dilemma or problem?" He knew the answer to this as he had seen her do it before. It amused him, simultaneously increasing his admiration of her.

Her eyes flickered as if searching for an internal memory. "I write it down."

"Yes, you write it down. You state your issue, and methodically plot the steps needed to solve the issue. It is really quite remarkable how your mind works." He was serious but he allowed a playful smile to grace his face and brighten his lips. She smiled back, playfully hitting his shoulder.

"My mind only works on occasion," she laughed, "but yes, I do like to see things written; it makes more sense to me somehow. I can visualize the steps and once in order they are easier to follow. Even just randomly jotting things down helps me to remember and organize. Like writing a book, some of it is just winging it but on a foundation of concrete facts and ideas." Seeing a look that she took as incomprehension she shyly looked down at her hands. "Sorry, that all sounds idiotic, doesn't it?"

"Ha, Adrianna, you are amazing. No one alive or dead could ever say you were an idiot. Your intelligence is only surpassed by your beauty, my dear." Without another thought, he leaned down to touch her open, soft lips. This time the energy was stronger, almost burning as their lips locked in a harmonized, intense dance. The kiss was long, deep, and filled with more passion than either of them expected or had ever experienced before. Minutes oozed by, their lips never ceasing the connection they made so willingly. Jonathan wanted to touch every inch of her skin, desire building and ablaze inside, a volcano before eruption, a tidal wave growing and peaking, ready to crash and flood. She reciprocated with an equal need, gripping the back of his head, drawing him closer and deeper into the lustful oblivion they had only touched on before.

Their bodies moving in sync, feeling blind, driven by a passion strong with desperation as if each touch would bring an end to their fear, shed light in their dark corners, and give freedom. Before either of them realized, Jonathan's, grey silk t-shirt was on the floor, leaving his hard-chiselled chest bare and enticing. Standing in only jeans, to Adrianna he looked like a god. His fingers graced her soft face before moving down to unbutton her light blue and grey plaid shirt, revealing her milky-white, flawless shoulders. Sliding her shirt down her arms, caressing her velvet skin, his breath caught in his throat when her shirt fell. He was terrified to gaze upon her slender waist, her small but full breasts hidden just enough for the imagination by a lacy white bra, her creamy skin hot and wet with desire, pale with perfection. He knew if he gazed too long he would be lost to the conscious world, controlled only by his searing longing to be one with her body and soul.

She shuddered lightly at the cool air battling the heat of her skin. Grazing her delicate hands across his strong chest and down to the button on his jeans, she released the hold of his pants and they slid slowly down his waist, buttocks, and thighs. Immediately she went to her own jeans, undoing the button quickly, and sliding them down until they too, dropped to the floor. Releasing her bare feet from the denim, she stood almost bare—alluring, radiant, and trembling with raw fervour.

Jonathan was lost, struggling uselessly to get enough air to clear his foggy mind. She was the most beautiful woman he had ever seen, beyond perfection, past all expectations he could have ever held. The angel before him was a dream, one taken from only the deepest fantasies of his mind, a vision too heavenly to conceive. Overcoming the fear of touching her, he tenderly traced her spine with his trembling fingers, stopping at the small of her back, drawing her body against his. Breathing in her natural sweet scent, filling his lungs with her aroma, his own breath coming fast and uneven, he convulsed dramatically when her hands traced his own spine and came up to entangle in his hair once again.

He knew he was past the point of no return, his control lost in the sweet abyss, and suspected she was at the same stage as he was. The passionate tension radiating from her body mixed easily with his own to create an almost unstoppable force. Their equal need drove them further until they were kissing aggressively, hands travelling over each part of their bodies, lost in the feel. Taking the next step towards complete and utter pleasure, they removed what little clothes they each still wore, embracing before slowly dropping to the floor, entangled together as one.

Neither of them heard the footsteps coming across the stone church floor and up to the little room off the back, though their bliss was broken when they heard the knock on the door. Scrambling as quickly as they could to clothe themselves once again, Jonathan answered when they were both decent, his boxers still on the floor, partially hidden under the desk next to Anna's Victoria's Secret bra.

"Who is it?" He asked, his voice husky and weak.

"Jonathan, is that you? It's Maria. May I speak to you for a moment?" He wondered with complete surprise how this woman knew where to find him, as he had only been in this room a few times before—with no one's knowledge, he thought. Maria? Though he remembered the name, he could not attach a face to the recollection. Pushing his mind through the erotic haze, he vaguely recalled meeting a short Chinese lady with some crazy death story, with Valerie, on an occasion or two before. But he had no idea what she would need of him.

He opened the door as Anna continued fixing her clothes, still bra-less, much to Jonathan's pleasure. The tiny woman stood, hands clasped in front of her looking rather curious to see Anna bent over trying to quickly grab the few remaining articles of clothing. Looking away with haste, she met eyes with Jonathan.

"Sir, I am so sorry to . . . ah, bother you . . . but Valerie sent me to look for you. She knew you were down this way. I am to bring you a most urgent message." She stammered with self-consciousness and obvious unease at the apparent situation she had just interrupted.

"Well, you have found me, Maria. What is this message?" he asked with impatient kindness. Adrianna stood holding the undergarments behind her back, waiting for the message as well.

"Well, Sir, there is a problem with Miss Dierdre. She has gone to sleep, and is not waking up. Valerie has asked that you and Miss Adrianna come right away." She turned as if to start leading the way, though the gesture was unnecessary as both Jonathan and Adrianna were out the door just after grabbing Shadow's note and stashing the unmentionables for a later retrieval.

Finding their way back with speed driven by panic and worry for Dierdre, both knowing that middle-worlder's do not sleep and should not be inactive, they ran faster.

Reaching the room only moments later, Jonathan saw Dierdre looking pallid, tiny, and asleep, lying on a huge, puffy black sofa. Her little, round, pixie face was soft and peaceful; no unnecessary, subtle, involuntary breath escaped her tiny body. She was as still as stone, wrapped lightly in an intricately quilted blanket. Valerie sat just feet away, ramrod straight in a small office-type chair. She seemed to have aged ten years. Signs of fatigue and worry were plain on her face as she sat staring at the lifeless little girl.

Jonathan felt anger at the woman but kept it in check to speak to her. After their last meeting, when her arrogant words had fired Jonathan's temper, he knew it would not take much to set him off. Though credit was due: she had sent for him and Adrianna; maybe it was a peace offering, he thought.

"Valerie, tell me what happened."

She was silent for a moment, gathering strength to tell the short tale. "I wish I knew. I left after . . . after we spoke earlier in the hall. She must have gone over to the table to draw—as you see, there is a picture—though I found her on the floor beside the chair. I do not know what is wrong with her, or if she is lost to us." For the first time, Jonathan saw tears in Valerie's eyes; he had almost thought her to be incapable of expressing sadness.

"She hasn't moved, although her facial expressions change. Sometimes she frowns, sometimes she smiles."

"How long has she been like this?" He asked, knowing there was no real measure. There were no clocks, and no one here gauged time in minutes or hours. It was a very difficult habit to give up.

"I was only gone for a very short time." There was no defence in her tone, only sadness and the underlying regret that maybe she could have helped Dierdre if she hadn't left her alone. Jonathan knew her guilt was enough; he didn't need to further her torment. A feeling of respect for the woman washed over him again and he forgave the conversation they had last had. Her good side, which she kept hidden, the side of caring, friendship, and even a maternal instinct was now clear in her eyes as she stared once again at the unconscious child.

"I am sure that there was nothing you could have done to stop this, Valerie. Everything that is happening lately has been beyond anyone's control. We will sort this out if we all work together." It was Adrianna who spoke to Valerie, with compassion and confidence in her melodic voice, the earlier pessimism gone for the moment with the need of her strength. To Jonathan there seemed no end to her courage and inner power. She gave him buoyancy, and he was glad she didn't know that.

Valerie stared at Adrianna in amazement, and Jonathan immediately worried that the old, reliably tactless Valerie would rear her ugly head, but to his surprise she softened, some of the rigidity leaving her tense shoulders.

"Thank you, Adrianna, for your beliefs and optimism. I don't know if you're right but I would sure like to stick around to find out." As the words came out of her mouth, she lifted her hand slowly towards her face, as if seeing it for the first time. It was shaking. No one spoke though there seemed to be much to say. The first to break the deafening silence was a very shaky Valerie.

"I believe I owe you each an apology. There is no excuse for my behaviour on our last meeting but I must say that I saw things that upset me." Her voice trailed off as if she was reluctant to divulge what had given her cause to act that way.

"Everyone is under a lot of pressure here, Valerie. I am sure you had your reasons, although I would not mind knowing what they were," Jonathan said slowly with just a hint of irritation. His earlier anger had been long erased by his worry for Dierdre.

"I saw someone trying to take Dierdre, but not physically. I know that doesn't make sense but someone was trying to control her. And then there are the whispers of those who are speaking to the others about not trusting you and Adrianna. I didn't know what to think. I know that this little girl is very important and I just wanted to protect her. Someone is trying to sabotage us and I . . . ." She didn't finish her sentence for fear it would upset both Jonathan and Adrianna to know she had doubted them.

Jonathan's retort was like the snap of a whip. "But you have never wanted anything to do with Dierdre. You have shied away from her; why now do you think she is so special?"

"I knew Dierdre before you, Jonathan. I touched her once, wanting to befriend her. I felt a pain that I had never experienced before and, well, I lost my sight for a while after that. So, I avoided her. I know she is special because Shadow told me a while ago that I should look out for her, that she is a "seer" too in her own way." Her explanation was complete and settled Jonathan's mind. He knew she was upset at losing Shadow and the others, and that she was also probably very worried about her own stability, having been here almost as long as some of the elders. The mention of Shadow recalled the note.

"Valerie, we have a letter from Shadow I think you should read," Jonathan stated as he passed the folded note to the woman. As she read with fascination and surprise, Adrianna went over to sit beside Dierdre. Smoothing the hair from her soft forehead, wishing she would open her green eyes and shed her natural light into the dark situation. When her hand fell still on the little girl's head a flood of energy, stronger than before, burned through Adrianna's hand and arm. It coursed through her body, leaving a gentle electrical sensation in every cell it passed. Anna felt alert and alive as she anticipated the visions that were sure to follow.

With conviction, she looked up at Valerie and Jonathan.

"This little angel is going to tell us what we need to know. Grab me a pen and paper."

9

Suspicions and the letter

I didn't know why I knew Dierdre would pass on to me some valuable information, but I believed that fact as strongly as the knowledge that I had died. Why she was no longer conscious I didn't know, but she would communicate with me—that much I knew. Jonathan passed me a large writing tablet and pen as he and Valerie both drew up chairs to be closer to us. Looking at me, he placed his hand on top of Dierdre's, and to my surprise Valerie did the same.

"Do you have a connection with her right now, Adrianna? Can you hear her thoughts? Do you know what is wrong with her?" Valerie asked, curious about my previous comment. (I wanted to say "here's your sign" but didn't want to be rude.)

"Well, no, not exactly. Dierdre sort of puts visions in my head when I touch her. Sometimes it is just pictures, other times it is full scenes drawn out like watching a mini video. She also makes things clearer for me, giving me her thoughts and knowledge as if it came from my own head. I somehow "know" things when I feel the current that passes from her to me." I stopped for a moment to allow this to sink in and waited for any questions. They had none. The shock and realization of what this could mean had them silent for the time being.

"It really is quite amazing what she portrays to me. I feel a current at first, like touching a live wire, then that subsides a bit and the images or thoughts start. I haven't seen anything yet, though the current is fairly strong."

I was worried that I wouldn't, that maybe she could not show me anything while she was in this state, like a coma. I wanted to help her, to find out why she was like this and if and when she would come out of it. Just when my fear hit its peak, I began to see an image in my mind. I decided to write down everything I saw, to hopefully later make sense of it all with the help of Jonathan and Valerie. I touched the paper with the tip of my pen.

A slight smile began to form at the corners of Dierdre's little lips. I knew I was doing the right thing, and I also knew that from whatever power this amazing little girl had she would give me as much as she could. I felt her fear, panic, and determination as I felt my own emotions of gratitude, affection, and amazement.

My hand started writing as the images became clearer in my mind. I felt the atmosphere like I had been transported to the place in Dierdre's mental state.

"Two women standing close, talking, at the edge of a cliff overlooking the ocean. It is dark, and there is an almost-full moon in the sky but no stars. The ground is hard and rough, and the air is cold. I can't see their faces. One is very tall and slimmer, with long blonde hair, the other short and large. I can't hear their words. I can't get closer. It is fading now." I stopped writing and noticed my audience was reverently reading my barely legible scribbles. Feeling slightly embarrassed I shyly looked into Jonathan's amazed eyes.

"I can't see any more, or who they are, but it feels bad—wrong somehow, Dierdre doesn't like these women. She is retracting. I'm sorry."

"No Anna, don't be sorry. This is wonderful. Maybe we can find out what is happening to her, relax please." His soothing voice re-vamped my earlier zest and I continued concentrating on the current still humming throughout my body. This time the image was of this very room. I could see the doorway and hear voices. Afraid I would not remember it correctly, I began to write it down again.

"I am sitting on the chair, up at the table." I pointed to the small wooden table that Dierdre had been sitting on before collapsing. "I have a pencil and I want to draw a picture. It is a scary picture and I am afraid." I shuddered with fear at the yet-unformed picture in my head, my hands shaking so much that the pen scribbled violently on the page. I passed the paper and pen to Jonathan. He took them with surprise.

"I cannot write anymore, Jonathan. Can you please write down what I say?" He nodded his head, worry displayed plainly on his face. Valerie sat almost as still as Dierdre, watching intently.

As I resumed my internal focus the image formed clear and sharp. At first, it was beautiful: the winged creature was large but dark, wings spreading open wide, black hair flowing gently over the shoulders. I gave a brief description of what I saw, shyly thinking it was yet another image of me as an angel.

"The angel has its head down, wearing a black, flowing dress; the feet are not touching the ground. Wait—its face . . . oh, it's not me." I gasped as the dark image became clearer in my mind. "It is a man, a young man I think. He is angry, mean. There are others, men in white robes, and they have wings too. They are beautiful, strong, but afraid. The dark one is speaking, his voice is deep and cruel." I felt the fear of the white angels rolling off like tidal waves when his words became audible. I recited the speech as I heard it.

"You pathetic souls may be satisfied with what you have been given. You may protect these imbecilic mortals and you may bring messages to their ignorant minds, but I will not. I have power beyond that. I will leave your union but not be gone. Here I will stay over your shoulders, watching you and your minions. My influence will give and take away, will save and kill and will be at my discretion." Involuntary tears formed in my eyes as the emotions from the dark room soaked into my head and filled my soul. This dark angel had power, more than the others, but I could feel it was limited and he knew this. He used intimidation to control, and the others feared his strength but did not know of his own insecurities. My voice shaky, I continued.

"One of the white angels is speaking now—he seems to be a leader—and they all stare at him in awe, except the dark angel. They know they have lost but they do not fight him; he is untouchable, the most powerful." I began again, a little more confident.

"We are not for fate manipulation. We aid, we watch, we bring messages to those mortals who need them. We do not choose who lives or dies. We protect." His voice was surprisingly calm despite his fear. He was old, timeless, and wise beyond the dark angel.

"NO! You protect, I will have control. Do you forget, Michael, who brought you the infinities? Do you forget who created those realities for the throne of Eternal? I do not forget that I have given you—and you, Gabriel—your positions; without me there would have been no need for you, as there is no need for you now, and no need for this reality where those mortals are hiding."

"We stopped you once, Lucifer. We will stop you again. You are bound to the mortal world, this you know. Your powers will not remain; there is one who can destroy you."

"You will not send me back this time, Michael. I will stop those mortals from living again, I will destroy the infinities I gave and I will destroy your jewel." I stopped speaking to swallow and clear my throat. The vision remained clear but now frozen, as if my human habit (completely unnecessary, of course) took precedence over the coming rebuttal.

"Adrianna, what is happening? Who are they?" Jonathan's voice broke my concentration as a white fog began to cloud my mind. Then, as if it had never been there, the vision was gone. Closing my eyes, trying desperately to get it back, I then realized that I was not meant to see more. That was all Dierdre had to give me. How she had that much, I had no idea, but I knew that she would continue to give me pieces of the puzzle until we could form the big picture.

"It's gone, Jonathan, I am sorry." Keeping my eyes closed, a wave of utter exhaustion took me over, and I felt like I had been running and running through a thick mud. My arms, legs, and head were heavy, the visions and concentration required having taken its toll on my energy. Shivering, I wrapped my deadened arms around myself, trying to feel warm again from my internal cold.

I opened my weary eyes slowly just as Valerie draped a soft, grey, crocheted blanket around my slumped shoulders. I muttered "thanks" and closed my eyes once more. Knowing I needed to tell them both that I couldn't continue, that I must rest before going on, didn't help the words form. My lips parted but no sound escaped. Strong arms and a smell of spice met me at the same time that I felt myself lean over, guiding my body back on the little sofa. My feet were raised up without my help, my head now on a soft, fluffy pillow that smelled strangely of fresh cut grass.

I knew I wasn't asleep, I could hear everything around me: the scrape of chairs being moved, the light wind of bodies walking away from me, the subtle sound of Jonathan and Valerie's voices far across the room. I didn't care what they were saying; what little energy I had left in my body was being used to concentrate on not touching Dierdre's skin so that I would not be immersed into another sea of images before I had time to recuperate from the last one. What I did notice from beside me, to my surprised delight, was that Dierdre's breathing had resumed being fluent. Even knowing we did not need air in our lungs (dead people don't have a life to sustain, ya know), and that breathing was not a viable sign of the condition of the being, I knew that Dierdre's breathing again was a sign that she was coming closer to being out of whatever comatose state she was in. Although I felt inside it would not be soon, I knew somehow that she would return and smile at us once again.

This realization made me laugh to myself. I rather enjoyed having these enhanced, assured feelings coursing through me, especially having come from a life of so much uncertainty. I lay still, silently revelling in the intricacies of my new-found and growing powers, and praying that I could expand them in time to be sure about other things. If Dierdre would be okay, did that mean we all would? Or was Valerie right when she said some would remain though not all? Will Dierdre stay on, mute, in this in-between existence, while the rest of us move forward to unknown lives, or disappear into oblivion? I wanted the answers to these questions, or a clue to which way this would all go. I decided before drifting a little deeper that I would just think about each step as it came, concentrating on Dierdre and my connection and what it could provide; the rest would come our way...I just hoped I would know what to do with it all when it did.

* * * * *

"She is exhausted, Jonathan. Let her rest, we must talk anyway." Valerie sat in the pink one of a different set of chairs, far across the room from the resting angels. Jonathan sat too, agreeing that they had to talk and Anna did desperately need to rest. He had never seen her so tired. Her normally exuberant, animated, colourful face had turned an ashen grey color, like death had come again for her. Her body had trembled, looking weak, withered, and beaten. It took every ounce of strength he had to not reach out and take her fragile body in his arms, draw her tired form tight to his chest, and release his energy into her through her luscious lips. But he knew that wasn't what she needed, or what Valerie would deem acceptable.

Cursing himself for listening to reason, regretfully he knew he must let her be for now. Before turning his attention from the lovely girls back to Valerie, he thought for a moment he could see a light, deliberate rise and fall to Dierdre's chest. Smiling, he thought he knew somehow that Adrianna would find a way to bring her back—her talents and love were endless and magical.

"So, you have this letter?" Valerie spoke, shattering his thought process, and engaging him in a more pressing matter.

"Yes, we found it in a book that Shadow had read from to me. He must have known somehow that I would return to his room and take that book. I don't know how. But he did. As you will hear, he gives us a lot of messages in this." Jonathan took the folded paper that Valerie had been holding out to him before he had noticed. Skimming through the lines to pick out the key points to comment on, he couldn't help but smile at Shadow's script. He never really had known if Shadow had been more impressed or irritated with him, though he suspected it was always a mixture of the two. His father having never been there throughout his upbringing made him feel like he was always missing that aspect of life, where a father shows his son he is proud of his progression into manhood, no matter how rocky the road was getting there.

Silently he wished that he had had a father to show him the way; perhaps then he wouldn't have made such drastic mistakes. In the end, however, he had done just as his father had . . . abandoned them all. Shaking the memory from his mind did not remove the guilt from his head, though it sat in the deep recesses until it would be called upon again to torture him further. Resuming his reading without the earlier pride, but still with the growing respect, he cited the first passage that had the most meaning to him.

"Shadow knew how special Adrianna is; he knew all along she would be the one. But he had no faith that we would prevail. He knew, though, that if there is a way, she is it." He glanced, equally worried and awestruck, at the woman who had stolen his heart and entered the hearts of so many others. She was the one, for everything.

"Here he tells us to read his books. I didn't realize it until just now but he has read to me from those before. They are about the war of the angels, and Dierdre's vision. She, I have no clue how, knows in that precious little head of hers the beginning and the end. She gave Adrianna what she had, and now we need to figure out the rest. What do you know about this?" Praying she knew more than he did, he now thought back and regretted not listening more intently to Shadow's stories. But he, in his weak defence, had thought that was all they were: stories, fables, the ramblings of some old man who wanted to write something to begin time, something to give glory to the highly-regarded word "angel."

"I have read those volumes, much time ago. I know of Michael, Gabriel, and Lucifer. The war of the angels was astounding as a tale and of course is hearsay at best. If it is true, and if the vision that Adrianna depicted for us has taken place, then the Dark Angel has returned and seeks to destroy us. He is behind what is happening. I will read the books again and try to gain more insight, but I fear they will not reveal the answers we need. If it is the dark angel Lucifer, then I have no conceivable notion of how we would defend ourselves against him, let alone prevent anything he has set out to accomplish. If our eternal demise is his order, then I believe we are doomed, Jonathan." Everything was silent. Neither knew what to say. Was there a positive, hopeful phrase? Only Adrianna could give light to such dark times.

"Lucifer? As in the devil? The little red guy with horns and a pitchfork? You can't be serious?" Adrianna's voice reached them with power, reverberating with the utter disbelief they all felt about the newly confirmed nemesis. She had made her way over to the conversing duo, and was standing tense and scared. They had not known she could hear them speaking, both now feeling uncomfortable at being overheard and not having had the chance to explain this to her before revealing the worst. Looking at each other in sorrow, they both turned to her terror-stricken face.

Jonathan knew Anna had taken the full responsibility of their fate on her shoulders. She would rather die a thousand painful deaths than fail them all. Knowing this as sure as he knew his own name sent shivers of sadness for her down his spine. The gravity of their enemy weighed down on her spirit, hope and determination. Anna had a difficult time absorbing the fact that she was the chosen one to help them save their world. Believing she was a plain, ordinary, fallible mortal, who did nothing special in life or death, kept her confidence at a lower level than it needed to be, or than it should be. Jonathan knew that, no matter the adversary, Adrianna's pure soul, compassion, inner strength, and wisdom beyond most would put up a fight to conquer any adversary of the ages.

Valerie spoke first, stopping Jonathan with his mouth open and words unspoken. "I'm sorry you heard that, Anna, but please, you must listen. I know it sounds like we cannot battle, we cannot win, but there is always hope—you have taught us all that." The smile Jonathan gave her, as if to prove Valerie's words, was filled to the brim with sincerity and gratitude that surprised Anna immediately. Valerie began again without missing a beat.

"Lucifer was an Angel, the most highly regarded angel, whose job was to discover and create worlds or realities, and he actually created this world. When he realized that humans had been given the ability to reincarnate, and in turn gain knowledge through lifetimes, he felt threatened and became angry. He wanted absolute power over all immortals, and felt that they must always be ordinary. So, he sought to destroy all who opposed him. He was beaten by Michael and Gabriel, the archangels from Dierdre's vision, and sent to the mortal world, but not before he had gained quite a following. Many wished to follow him to whatever depths he may travel. No one knew what happened to him: if he became human and reincarnated over and over, I can certainly think of a few humans I have known personally and heard of that sure could be compared to the devil." Valerie spoke with kindness and precision, hoping to both calm and educate Anna, as Jonathan listened intently at her side, now holding her cold hand.

"Fact is, if Dierdre's vision is right, then he is back and wants to annihilate our world." Jonathan saw Anna cringe at Valerie's last words. Holding her hand tightly, he knew he must say something to soften the blows to her poise and hope.

"It's a miracle, but Valerie may be right," he said, quickly giving her a look to assure her he meant no disrespect. "However, Lucifer has not come across you before, Adrianna. He was beaten before and will be beaten again. We may all have reincarnated and ended up here for a reason, fully equipped with powers and talents. He knows nothing of what you, I, Dierdre or Valerie can do. I bet he doesn't expect a battle: wiping some of us out as we were unprepared was his way of playing his game. But it is our move next." The sureness in his speech did not reflect the inner turmoil now raging deep inside him. He wanted intensely to believe his own promises, but he knew he was transparent.

Adrianna smiled at Jonathan, her unbridled beauty and kindness never so evident. He knew she was putting on a brave face; as usual, her desire to help and protect others was above all else. Still pale and shaky, she began to look a little better than she did before she lay beside Dierdre.

"Ok, this Lucifer is responsible for what has happened to the lost ones, and he will continue to try to destroy all that is here because he is jealous that we are talented and can move on? A control freak that needs to dominate to feel like a man. I'm now convinced it is just the male species that is that immature, angels or not," she stated with the slightest hint of humour in her vastly deep chocolate eyes. As usual, her strength and ability to simplify the most complex situations was astonishing to behold. Even Valerie could not stifle her grin as Jonathan outwardly laughed at Anna's synopsis.

"Yes, I guess that is it in a nutshell," Valerie commented, still smiling despite the worry that was clear in her eyes.

Adrianna straightened her shoulders as a warm flush appeared on her cheeks. The power she held flowed through her and radiated heat and energy all around her.

"So, the little red guy, complete with horns, tail, and little, toy pitchfork needs to be put back into his place, and we are just the quartet to do it.

I guess we had better get busy." Taking Valerie's hand and tightening her hold on Jonathan's, she allowed her positive energy to surge into her comrades, giving more than just strength: she gave courage, enlightenment, and spirit. They both responded in shock at the current running through them as Valerie took Jonathan's hand to complete the circle.

Valerie spoke first, her voice strong, and sure: "I will go and read as much as I can to gain more information on how he was defeated before and how he is bound to the mortal world. I will come back soon."

Jonathan was next to speak, as his courage reached a peak: "And I will go speak to the others in the common room and get a feel for what everyone is thinking. I will also track down anyone who might be able to help us in any way."

Following suit, her resolve complete and acute, Adrianna spoke: "I will sit with Dierdre a bit longer to see if she has any more to share with me."

Valerie left first, without a backward glance, seemingly charged with their powerful encounter and with the mission at hand. Jonathan hesitated for a moment with a hint of reserve on his radiant face. Words that would just not come were sitting on his tongue as he gazed at Anna. Feeling her rush of anticipation at what she might see when she again touched the little angel lying so peacefully on the sofa, he worried that she shouldn't be alone, but, urged by the intuition that he was needed elsewhere, he resolved to leave her for as short a time as possible.   Touching her hand once again caused no shock but pleasant warmth through his oddly cold hand, travelling on to encompass his being. The warmth that spread was not the same electrical charge from before, or like any of the other times they had touched and felt the strange sensations. This was from the heart. His heart, her heart. The phenomenon of affection, of aspiration, of love.

"I don't want to leave you." The words and apparent emotion behind them shocked Jonathan as he heard himself speak. He had not meant to start with negativity or say that at all. Again, he marvelled at how loose his tongue became when he allowed Adrianna to mesmerize his thoughts and mind. Her fine, dark brows drew together slightly but her lips smiled in reciprocation.

Always, her kindness and incredible understanding proved to be more than Jonathan could ever hope to have and brought with it the sad emotion of worthlessness in her genuine presence. Wondering how the cocky boy from his past had been so wonderfully humbled in such a short time brought a smile of bewilderment to his own lips. He did not mind that she misunderstood the reason for this smile, as it would only hurt her to know what he had really been thinking. She disliked it when he brought himself down in comparison to her angelic qualities, as she did not believe (no matter how convincing the arguments) she possessed any special attributes at all.

"I mean to say, I wish that I didn't have to be away from you right now, or ever, really, but I must go for now." He blushed at his forthrightness about his emotions, still being new to the expression or even the acceptance of such deep passion and love. Never in his dreams had he envisioned what had transpired between them and what potential there was left for more than even the heavens could conceive.

"Yes, sweet Jonathan, you must go, and please do not worry about me, it will cloud your judgement and narrow your intuition. Go and seek out those who can help us, calm the fears that must be running rampant in their minds right now. I know you, if not anyone else, can bring them order and hope." Smiling, she laid her warm hands on his arms, and then slowly moved into his arms. He stood dumbfounded for a moment at her confident words of wisdom, her gentle touch, and warm embrace. How had she come to him—him of all darker souls, him who didn't deserve love and compassion? But she came and opened her own vast heart to him, gave her gifts without wanting anything in return, so willingly, openly, and freely. He would never deserve her.

Returning the embrace, hiding his tortured expression from her intuitive chocolate eyes, he breathed in her engaging scent and held it inside, allowing the aroma to flood him, soak into his very soul.

Knowing if he did not break away at that instant he would not ever let go, he stepped back quickly, gauging the expression on her face. As usual, he knew, she felt the same building need. Turning from him, giving him his leave, she went to sit beside Dierdre to delve into the mystical little mind of cryptic answers. He left as Valerie had, without looking back.

Jonathan forced his mind onto the task at hand, allowing his passion for the beautiful woman, left behind for now, to sit pleasantly in the back of his mind as he entered the common room. Much to his surprise, it was nearly empty. A few people he knew well and a few he did not were standing around one of the tables. Approaching slowly, without drawing attention, he grabbed a chair nearby and sat. They all turned to stare at him as if he had three heads and six eyes. It was then he saw the little stone figure on the table. It was then he realized that someone had already been to do what he had planned to do, but in a much different way. It was then he knew he was in serious trouble.

10

Violence and Faith

I knew if I did not stop thinking about Jonathan and how amazing his arms felt around me, I would never get close to figuring any of this out. But every time we touched, even without the added buzz of electric shock (now that was freaking crazy and hugely pleasing), I didn't want to let go. I wanted to touch him, explore him, like we had in Shadow's office. (Omigod, I'm blushing again.) I couldn't believe I was thinking these erotic thoughts when the world as I knew it was dying and my new little friend had fallen into a state I couldn't seem to snap her out of. Apparently, the teenage hormones that had started rushing around puberty didn't exactly die with my body (go figure).

Knowing any happy or unhappy, appropriate or downright wrong thoughts I could be thinking about him would go away as soon as one of Dierdre's visions enveloped my mind, I quickly grabbed her hand, holding it tight to my heart. Although her little body didn't move, I felt a tremble from within her and within me. The link re-established fast, as if she had been waiting for me to re-connect, images ready to flash like YouTube clips of horror movies.

When the first vision appeared, I was not at all surprised that it was fitting to my morbid thoughts of bad thriller films (I had seen way too many of those), though I was very surprised to recognize the place immediately. In a split second, I had completely left my body, sitting on the little sofa next to the soft, warm child, and entered a room I had known very well for many years. It was a room that, without exception, gave me a bad feeling each time I entered, a room that housed a monster. This time it was not Dierdre's monster—it was one from my own past, my own life.

The room was dark, as usual, curtains drawn to dispel any natural sunlight that might possibly want to enter this evil room. Through the scant beams of light that managed to invade, a thick almost odorous dust sifted through the air from many years of not being cleaned away by a vacuum, a duster, or any other cleaning agents. The monster stood in front of his filthy, stained dark-grey couch and cluttered coffee table. There wasn't a clean surface to be found. The tiny area could only contain one couch, one tattered brown chair, a small TV on a cheap metal rack and a bookshelf that held a VCR, DVD player, and more porn mags than an adult store. An overfilled ashtray lay among papers on the side table with the broken Marilyn Monroe-figure lamp; another was spilled on the dusty, brown, old shag rug. Not three feet from the monster was his son, wearing a look of hatred that made the monster's squinted face look almost pleasant.

The voice tore into me like a crack from a whip, leaving behind an open gash. I felt like I was bleeding, cold, and weak. It came loud, clear, and strong. I was there. In that hated room. The smelly torture dungeon. I could feel the air, taste the rankness, sense the hatred, and suffer the fear. But it was wrong. It wasn't like before. Was I confused? Not noticing the vibes correctly? I was almost sure the fearful vibes were coming from the monster himself, not the other figure standing—the usual victim, my beloved friend Willie. But no, I was not wrong.

The vibe of power, the sureness of force, strength, and intimidation was coming from the opposite opponent, the one who would normally cower, tremble, and bleed. This time the monster was the wounded. It was hard to conceive, but the vision and its components were real and impossible to ignore. As my acceptance of this strange turn of events became clearer, so did the scene before me. The props came into play.

There in Willie's hands was a bat, a baseball bat, wooden and well used. I had held that bat myself many times before as Willie pitched underhand softballs at me from a short distance to train my weak skills at one of his favourite sports. He had used that very bat to slam a ball over the Crosby fence and into the window of their garden shed, later to receive a serious whopping from his father, so many eons ago.

Now the bat, sprayed on the end with what I was sure was blood, was aimed at his father's head, ready to strike against his skull as it had just moments before. The monster had his hands up in defence, one holding the side of his face with blood seeping through his fingers. He wavered and trembled from the injury or drugs, I could not be sure—likely a combination of the two.

"I told you I would make you pay, old man. You drove my mother away, my brother, my life. You deserve to die for all the shit you have done. I have nothing because of you. Tell me why I should not kill you now?" It was the voice I had listened to for so long, but not a voice I had ever heard before, though I had caught a hint of it on the last visit to Willie . . . . The harsh brutality was new, the softness gone. The positive, shed-light-on-this-shit tone as dead as I was. I did not know this man, this mad, out-of-control boy whom I loved. I could not fathom the scene before me, although I had seen it in a million movies, read it in tons of books, heard it on dozens of news reports: the victim gets revenge, years of torture built up then released in a fit of rage.

But Willie wasn't like the characters in the books and movies, Willie wasn't the one to hit the headlines, Willie was saving money to leave the hell he grew up in and make a life for himself far, far away. My Willie wouldn't take a bat to his father, no matter what horrid things the man had done to him.

This Willie had an edge I didn't know existed, a hatred that went beyond what I had known. His bloodshot eyes, wavering stance, and pallid skin told me more than I wanted to admit. He had taken on more than the violence that his father had given. He had taken on the drug habit his father had lost himself in, which had destroyed him as a person, a husband, and father.

Willie, my best friend, the only boy I had ever loved (besides Jonathan), had turned into the beginnings of the very monster he was set on destroying.

"William, put that bat down. I gave you that for your birthday, remember? You were . . . ummm . . . you were fourteen that year." His father, the snivelling, five-foot-ten, overweight man, stammered his words through tears and blood.

"No, you stupid fuck. I wasn't fourteen and you didn't give it to me. I was ten. I found it in a ditch along the highway and you accused me of stealing it, then you broke two of my ribs, telling me that stealing was for weak idiots that deserved to be beaten." Willie screamed these words, his voice cracking and shaking with rage. I remembered that day. The happiness he had felt to have his very own bat, despite the bandaged ribs and searing pain he must have felt with each breath he took. We had played ball so many times. Willie had often dreamed of being a pro ball player, knowing it would never be so, knowing he had few options in life, but he used to have hope. He wanted to become something, to use his talents in fixing electrical devices, his ear for music, his gift for drawing. There used to be a special substance inside him that predicted and assured victory, to rise above his past, to be someone other than the torn, beaten, sad, scared little boy.

That was the Willie I knew and loved. The boy I watched grow and become strong, the sweet, adorable man I saw him become in my own dreams. I even wondered if eventually we would fall in love, if the feelings I had of friendship, protectiveness, and compassion would evolve into something deeper, something more romantic and intimate. Although I did not feel that way about Willie, I was not averse to the idea of my feelings changing and becoming much more.

But now things had gone wrong. I had died. Left him alone. I hadn't forgotten him, but I had moved on with the existence I had now and fallen in love with a boy I hardly knew. Willie had gone down the wrong path, changed, warped, and potentially destroyed his own future. He had become his father, the very man who had destroyed his childhood. Astonishment blanketed my thoughts, coupled with the agony of my guilt; I felt a vague trace of the hot tears burning down my cheeks. What had I done? What could I do? Was there any way to stop this roller coaster of hatred speeding in a downward spiral? I wished I knew if this scene had happened, was happening still, or had yet to transpire; and if it was in progress, could I intervene before it escalated to an unknown but definitely bad end?

"William, I was always hard on you for your own good. You needed to be taught, you were so weak, just like your mother. I had no choice but to toughen you up. Now look, you are strong, I'm fucken bleeding. You have me, boy. Now let's end this. Just leave—that's what you have wanted to do, planning with your stupid little friends."

Switching from fear to anger, Willie's father stepped forward a few steps, closing the gap between him and his son. The blood was coagulating on his red face, the head wound looking serious and not so unlike many of Willie's past wounds; the scene was like a horror movie. I knew he was as afraid as his own boy had been so many times before. His damaged, stifled pride bubbled to the surface of his words, trying desperately to gain back some of the lost control. Failing to intimidate the child before him, his fear grew, threatening his mental status.

"You should be dead, and you might be yet. I will leave, but not until I have shown you what you have really taught me and how tough I really am."

The smile that I had never seen before, playing at the corners of Willie's full lips, contradicted the cold rage in his vacant, soulless eyes. I watched the bat twitch is his unrestrained hands, the rage still building inside his tortured being and the desire for retribution flashing across his hard face. I wanted desperately to stop him, to tell him that this wasn't the way, that he could just pack his stuff and leave, as his father had said. He could still have a life outside of what he had known. With help, time, and peace, he could heal, learn to forgive, and rise above the dirt he'd been buried in.

"You have done it, boy. I see you for who you are: you're just like me, you little asshole, without your stupid, dead little girlfriend you are dead like me. You can't kill me—I've always been dead."

What happened next, I knew somewhere inside, was inevitable, the predictable conclusion to such a horrible encounter. Nothing good could come from the two forces colliding in front of me. The action was in slow motion, separate but conjoined decisions made simultaneously, my own included. I reached out with my mind and every ounce of strength, every semblance of power I could muster, and prayed for an intervention, a chance occurrence to prevent the next expected action from finalizing the fate of the two men. Feeling a surge through my body told me something was happening, just as I heard loud and clear the words that told me I was too late, not strong enough to prevent this tragedy.

"Don't you talk about Adrianna! You're right, you are a fucken dead man." My best friend's words died with the deafening sound of bone, skin, and blood being crushed and destroyed with the force only years of hatred could produce. I watched the side of the monster's head cave in at a grotesque angle and form. I watched him crumple to the dirty carpet, never to strike again the innocent boy that had just died with him. I watched my childhood love, my best friend, die a worse death than I could imagine. I watched Willie's soul leave his body as his father took his last breath.

Whether it was my tears that began the clouding of the vision from Dierdre or the fact I felt like my heart had just exploded in my chest, I didn't know, but the ugly scene before me faded and I was left terrorized, pain wracking my body in torturous waves and tears flowing, blinding my sight. Knowing I did not need to breathe to exist did not dispel the fright I felt at the band that was now wrapped mercilessly around my lungs and body. Nothing I had ever witnessed before, not even my own death, came close to the terror I felt now.

Comprehension was distorted by hysteria. Witnessing my best friend committing such a brutal murder of his own father topped the list of the worst things to go through, and coming from a dead girl that was pretty bad. My body was jerking spastically with each gut-wrenching sob that busted from me. I was so very glad I was alone, that no one knew what I had just endured, that no one could see me folding in on myself in a suffering even I could not describe. The holes in my insides from the daggers of pain I had felt before Jonathan had found a way to half-heal them, were wide open now, raw, burning, tearing bigger with each re-enactment my mind tortured me with. No horror movie could compare—Stephen King could not articulate the fierce emotions ripping me apart.

Trying to gather my wits and rid my mind of the replay, I looked over to see Dierdre frowning with a look of sorrowful ache that I knew almost equalled my own. She had seen it all. A part of me felt comforted that I was not completely alone, but I was horrified to know that this little angel had to witness such vicious hate, and my reaction. Her face also showed an emotion that sent a disturbing feeling through my hand, which clutched hers still. She had understood. She had related to Willie and his hatred, she revelled in the vengeance, and felt satisfaction to be privy to the monster's last breath.

This realization almost wounded me more than what had transpired. Had she been where Willie had been? Did this vision bring out her own past, with her own monster from the vision she had shown me before? My questions swirled in my mind, distorted, painful, and incomplete. I wanted the answers as much as I didn't want to even have the questions. Wishing I could hide, dissolve, float thoughtlessly into an oblivion where fathers did not hurt children, where best friends did not commit heinous acts, where happiness was everyone's strongest pleasure and emotion, where your enemies could be beaten without sacrifice and loss.

Dierdre's eyes fluttered and I knew she felt the same. She had wished my wishes and held my dreams once, but now she knew that it wasn't that way and wouldn't be; we could all die and still suffer, but not to dwell, but to relish in what good there is and how wonderful it could be. To bring out happiness in pictures, smiles, and laughter, even if it was silent.

I wiped my slowing tears on the back of my sleeve and took a useless but relieving deep breath. It calmed me, and Dierdre calmed me. Her desire to see goodness, color, and beauty flowed into me; her desire to be happy and accepting pushed into my soul, clearing the residual hopelessness away and making a path for focus and insight. She would show me her past, the vision that would answer my questions regarding her, but not yet. She knew I had just been through something I would not easily recover from, if I ever did. I felt certain that if I lived a thousand years like Shadow, I would never forget what I had seen in that living room or the damage it had done permanently to my soul.

* * * * *

Quickly correcting his stunned posture into one of assurance, Jonathan's eyes narrowed into slits, his hands balled into fists. He began an ad-lib speech of anger that he hoped would have the effect that might very well save his mission into this already-diseased room. He was certain someone or some others had knowledge that would help them; he was also certain that Lucifer had followers here, possibly in this very room. Someone was giving the dark angel information and allowing him passage into this place. The devil would not work alone; he would have minions wherever he needed them to be to do his evil bidding. He could not think of who would aid a demon to destroy their very world, but he was sure he would find out.

"So, you are scared, despairing over the loss of loved ones and friends, and you choose to abandon your inner good to honour a devil, a creature intending to destroy us all? Is that what your cowardice has brought you to?" Coming on strong, he could see the shock, guilt, and avoidance in many of their eyes. He was right. They had been presented with a way out, a promise of life somewhere else, a false guarantee from the dark angel himself. He needed to think quickly and show no mercy.

Before he could continue, one of the biggest men he had ever seen in his life and death stepped forward to confront him. "Now wait a minute. You waltz in here, after ditching us, when you were supposed to give us some answers, and you call us cowards?" The ex-truck driver from Northern Ontario loomed over Jonathan, equalling his intensity, and stifling some of Jonathan's fictitious confidence. The fear he felt quickly converted to anger and strength. Words from Shadow's letter loomed in his mind. They must unite.

"I am working to save our world; I can't be here to hold everyone's hand. You must be strong on your own, each of you, standing together. We are very close to knowing what is behind this and putting a stop to it. There is no other option but for all of us to work together, which is why I am here now."

Silence followed his speech, guilt remaining in those who would meet his eyes. Jonathan had always been the one to deliver messages to the middle-worlder's, though he had spent little time getting to know most of the inhabitants, instead having simply listened to and almost immediately forgotten their stories and even their names. But he was charismatic; people listened to his words, his jokes, his tales, and his direction. Valerie and Shadow would ask him to give all the speeches and talks and he would do it to pass the time or just to be the spotlight for a while.

As he looked around the room he realized even now that he could remember few names and even fewer of their past lives and deaths. His own guilt blinded his vision for a moment until he tucked it neatly away with the rest of the mounting remorse he felt at his ignorance and selfishness.

Knowing it was time to soften his words, he started much differently. "I know you all need more: more proof, more answers, more hope. I want to give it all to you. I want to tell you we have saved our world and the threat is gone but I can't, not yet. But I will, with your help. I would like to speak to each one of you individually and ask a few brief questions. That's all. But I need to know that none of you have made any decisions on your own first, so if you have please tell me now."

No one spoke at first, but he knew it was coming: one would start and the rest would follow, and they would tell him what he needed to know.

A tall, well-built man who looked remarkably like Clark Gable (no, that was impossible) stepped forward in the place of the trucker, who had, thankfully, retreated beside the little blonde woman standing meekly at his side.

"Maybe we have been given the answers we have asked for. Maybe we already know what we are going to do."

"No, Danny, wait—we haven't decided anything." A very attractive brunette woman close beside him grabbed his arm.

"Well, I say that I have."

"I haven't decided anything. I like it here and I want to wait for Jonathan," a gray-haired woman said.

All at once, they were talking: to Jonathan, to each other, and to no one in particular. Agreements, denials, fears, certainties, and angry remarks were flying around the room like toothpicks in a twister. Jonathan couldn't determine how many were content to wait for him, or how many had taken the advice of the devil's messenger, but what he did know is that he had to act fast or lose the majority. Without really thinking through his next words or the ramifications, he approached the group again.

"Listen! Listen to me. I see that you have been told by someone that you have another option, that we will not succeed and your only hope is to follow the advice of another. By the statue on that table, I can tell you that you will die if you believe the promises you have heard. This person who has told you that you must move on, or whatever has been directed, is trying to help our enemy. If you do not want to believe me, I understand, but will any of you doubt Adrianna?"

This brought shock to his audience quicker than any other words that had been spoken. He knew that everyone, not one excluded, believed

Adrianna to be an Angel. They had all heard the tales of her powers (even though some of the stories had been stretched a little), they had believed she was the only one who could bring hope, and they all adored and admired her beauty and strength.

Some had caught on to what had been transpiring between her and Dierdre, furthering their faith in the special abilities. But Jonathan had told them all that for Adrianna to help them they must leave her be, and must not ask her questions or beg for her help. He had done this to protect Adrianna before she found out who and what she was, and after, to keep her from being pressured and overwhelmed. Valerie had disagreed, thinking Adrianna could keep up the morale but Jonathan had been selfish, wanting to shield her and keep her to himself.

Now he realized they had probably lost their faith in the prophecy surrounding Anna and had given up on her abilities, not having seen any proof or even seeing her at all. He didn't want to expose her to the doubt, the fear, and the disbelief in the hearts of all the occupants of the lost world they shared, but he could not instil confidence into their souls alone. He knew the traitor had won many of them over and he knew winning all of them back might just be impossible. It might take a miracle to have each one of them back on their side, ready to fight the most spiritually evil adversary any of them would ever face. Adrianna was that miracle.

Addressing the crowd, who were still whispering amongst themselves, Jonathan felt his own belief shaken by their despair. "I want to hear you all but I cannot this way. I will speak to each of you, one at a time, and then bring to you one who will calm your fears. Please allow me to open your minds again. Not all is lost. But there is only one way, and that is for us to stick together and let no one else into our solid group, our family. The union we have built, each one of us, systematic, room by room, beautiful place by beautiful place. We can do this as one."

Murmurs of skepticism of defence and of division still flowed through the room but, even with the doubt, they still gathered slowly in a line behind the table and chairs. Jonathan, surprised but pleased, sat back down in the chair at the head of the table and motioned for the first friend to sit in the chair closest to his right. The others stayed at a respectful distance, talking quietly among themselves. From the corner of his eye, he noticed a few leaving for unknown destinations. This was a problem. There were some who were solid in their minds to rebel against Jonathan's lead and to follow the path of their new, treacherous foe. Keeping focused on the many that remained, he began to ask questions and gain information, worrying about Adrianna and what he would have to ask her to do.

11

Angels (good and evil)

I felt detached from my own physical being, much as I had in the beginning, and silently wondered if I was doing so to ease the pain that still held me. Though I felt clearer, calmer, and less shattered than I had a few moments before, I still needed relief from the toll the pandemonium of the last hours had taken on me. I wanted to rest but knew I could not. There was no time for a breather (haha—a great word from my alive life, so unnecessary now), only time for action. I had to find a way to reach Willie. I didn't even know if I could see him again and speak to him like I had not just witnessed him committing murder. If I found a way, gathered the resilience and power, could I face him?

Stopping the rush of despair before it completely engulfed me, I grabbed hold of Dierdre's right arm and hand. She felt amazingly warm, soft, and cuddly. This softened me immediately and I took a moment just to feel.

"Dierdre, I don't know if you can hear me, if my talking to you will help you at all, but if you can listen to me, please do. I need your help" (I sounded like such an imbecile).

"I know you are trying to guide me with your visions, but I don't know what I am supposed to do about Willie or anything else. We figured it out about Lucifer, but how am I or we supposed to defeat the devil himself? I am not what everyone thinks I am and I am scared I will be the reason we fail." Silent tears slid down my cheeks as I spoke to the unconscious angel before me. I felt no shock, no buzz, no anything from her skin but somehow, I knew she was hearing me. Something was telling me to keep talking to her, that she would collect all my words and compile the best answer she had.

As soon as I thought it, I knew it was true. Dierdre's thoughts came rolling into my mind as if they were my own, as if I were thinking them, though I knew they were from her. My questions were being answered one by one as best as the little girl could manage with my jumbled thoughts and rampant questions. I spoke the messages aloud, both for clarification and just to hear myself speak her silent words.

"I should not try to speak to Willie; it is too late for that. I could hurt him further. Time will show me Willie's fate and how it is connected to me. I need to focus on what is happening around me; Jonathan needs me more than I realize right now, and it is very important for me to be there for him and have the strength to do what he needs me to do." (Oh man, why does everyone need me?)

"It will not be just me that can defeat our enemy, but the union of a few who are pure, led by one. I must have faith, hope, and compassion even for those who do not prove their worthiness of that same compassion." I finished speaking as the last thoughts sifted through my mind, bringing with them more tears, burning my eyes, and a heaviness of burden in my already overfilled heart. 'I need you too,' I heard in my mind, though I did not speak it aloud. Dierdre needs me. She didn't mean for me to hear that thought but I did. She is frightened, alone, and just as confused as the rest of us, and she needs me too. I knew I must find a way to help her come out of whatever hold had taken her away from us. I needed to reach into her mind and search for the one answer I had not asked for. Not knowing if I had that ability to delve into her mind without her opening that usually one-way portal we shared, I sure was going to try. I feared she was weakened and my intrusion might harm her, but if I didn't try, I would never know how to save her. I wasn't even sure I could if I knew how.

My confidence had been damaged by the knowledge she had given me about Willie. I was not able to help him—it was too late. Had I acted before, could I have prevented the fate of my lost best friend and his father? I didn't think I could have. Whether it was my heart protecting itself by not allowing itself to believe I failed Willie, or not, I was almost certain that no matter what I might have done since I died, nothing would have changed the course of events that took place since.

Wanting it to be different didn't make it so.

I had so much information rolling around in my mind like wet clothes in a dryer. I wasn't sure if I could muster the focus I needed to attempt to push inside the little angel's mind for a clue, a vision, or a thought that might help me to unlock the mystery surrounding her comatose state, her inability to speak (minus her kick-ass way of communicating) or the reason for her existence here in the middle world.

I was just reaching for her little hand again when Valerie burst through the doorway in a rush. Her aged but handsome face was flushed red, which accentuated the dark circles around her deep green eyes. She must have been a striking woman, as she still had a beautiful air about her.

"Adrianna, I haven't had an informative vision in so long, but now I have. I don't know what to make of it, though—it is complex and unclear." She seemed to have lost her usual refined composure, as her words came fast and without the usual dignified delivery.

"There are others who are helping Lucifer. He is behind this; he means to destroy us and now he means to destroy you. But somehow, he cannot touch you; he fears you and I do not know why. He has a plan, and he has helpers to carry it through. They are here among us and Jonathan will speak to them, although he will not know what or who they are. You are the key, Adrianna, you really are. If I doubted you before, I do not doubt you now. He may not be able to defeat you—it is like you are protected—but don't feel settled in that fact, my dear, because he means to hurt you any way he can and he will not stop until he succeeds."

Her breath was coming in raspy jolts like she had run one hundred miles. She was elated to have had a vision, pleased that she could now believe in me even though I did not believe in myself, and terrified at the intensity she obviously felt at the information given to her by her special sight into the future.

"Valerie, I don't know what to say. We must go to Jonathan; he might be in danger. I don't know how I am special enough to be shielded from this awful sadistic angel but I sure hope you are right. I want to say I am the key, Valerie, but what if I am not?"

I feared the answer as much as the question but had to ask anyway. I wanted a distraction from my worry for Jonathan. He would be a bit put off by us two women barging into the common room to tell him he was in danger and we were there to save him. Oh no, that would not be good. I had moments to think of a better plan.

"Adrianna, Shadow and Hunter and the other elders knew of you when you were born. They did not know what threat would be posed to our world, as they could not conceive of a possible danger to us. But they knew when the danger came you would be the one to save us. It has been talked about for a very long time. When Jonathan took it upon himself to get to know you, the elders feared it could change the prophecy. Then when you died and Jonathan was to blame, some believed he had damaged your ability to be what they thought you to be. I doubted you too. I had to. It is the way I am, despite the visions I had that said otherwise. You are the key—you always have been."

Her words then came to me. I heard each one, filed them away in my memory, and then replayed them. Something she said struck me, hurt me, and scared me. She had let it slip: the secret I had caught myself on a few times before, but never completely held in my mind long enough for comprehension to clarify its reality. Jonathan had brought me here. Jonathan was to blame for my death. Jonathan had killed me?

I had asked him before, not so long ago in the place that reminded me of the Star Wars movie (I love the Ewoks), if he knew who had killed me; he said he believed he knew who had. He had even indicated, talking to Shadow, that he, himself had been the one to kill me, but I had instantly disregarded it, thinking that it was just his guilt causing his belief that he had. I believed then that it was possibly Shadow who had brought me here, or the elder Hunter, who had been with Jonathan in the park. I didn't actually believe he had done it.

I didn't understand how this could be possible. He wouldn't do that. But he did, somehow— I knew that now. It hurt me to know that I had lost all my friends, my family, and my life so early, and for what reason? Jonathan. I knew why he wanted me here. I knew why he took me away from the ones who needed me. I had already forgiven him for wanting me to be dead, to be here, but I still did not want to believe that he had actually killed me. (Yeah, twisted.) I didn't want to believe that it was true; my heart was so filled with him. His face, his touch, his smile, and his laugh were never far from my thoughts, my heart. The boy I loved, the man who held my heart had taken me from my world and brought me into a new reality I should not have had to know for many years.

My body felt cold and I could feel my face pale. I knew Valerie was still speaking to me, although I did not hear the words coming from her moving mouth. Valerie slowed, then ceased speaking as my new demeanour was made apparent.

"What's wrong? You have heard us tell you how special you are all along. This should not come as a shock to you now."

Her statement was stated as a question, as she obviously could not understand the thoughts flitting through my mind. I wanted to tell her. I wanted her to tell me it was not so, to retract her previous statements, void them from my mind record and bring me back to where I was before. My focus was lost. Knowing I needed to put this aside and trudge on did not help to ease my mind.

"I have upset you, Adrianna, and I am sorry, but you need to focus now. We have much to do. May I ask, though, what I said to cause you to . . . well . . . you look like you have just lost your best friend." The worry was plain on her mature face. She was trying without success to hide her irritation at my silence but I just couldn't find the words. She had no idea how correct she was. I had not only lost my old best friend in one final unforgivable act, but now I had also lost my new best friend.

Stop! Did I truly feel that way? That I had lost Jonathan? He may have killed me and brought me here before my time, but I knew why. He had been honest with me and I had already subconsciously forgiven him for the possibility that he might have had something to do with my death, so why was I having such a fit now, when it would do no good?

I knew Jonathan cared deeply for me. I felt that in his touch and saw it deep in his vibrant blue eyes; he would not have done anything to me to cause me pain without a reason. Believing my own feelings on the matter was more difficult than I wanted it to be. I knew deep down it was my own past, my mother's past, and my own insecurities that caused my distorted beliefs. I just couldn't trust in feelings, even, sometimes, my own. (I need therapy.)

My mind was whirling as Valerie's face grew stern. I knew I had to say something to ease her before she erupted with exasperation. But what could I say?

"Did Jonathan kill me because he knew I was needed, or because he wanted me here for himself?" Whoa, I hadn't really planned to say the first question that kept plaguing my mind. Crap! Too late. Valerie's face changed from irritation to surprise, then embarrassment at the realization that she had just revealed that fact to me, and then she softened to understanding.

"Adrianna, when I see your face right now as we stand, I know why Jonathan loves you and has loved you for longer than even he knows. I cannot answer your question with proof or enough conviction to settle your heart, but what I can tell you is that Jonathan loves you, he has always loved you, and you are needed here to save us. That is what I will help you do. We must go to Jonathan." Her words were strong, firm, kind, and full of truth. I could not deny that the hurt I felt was still there. The doubt surging through my heart had weakened but not yet disappeared. I would need to talk to Jonathan and take whatever he may say or not say. But now I had to face him, knowing what I knew, feeling how I felt, without the ability to ask him to clarify or to explain. I had to follow Valerie, as she had already started walking away from where we had stood together, both revealing too much.

I followed her down the hall I had travelled countless times, to the room I had entered almost as many times. Taking in no changes in my surrounding, feeling no vibes, sensing no waves to direct me into what we would face in the common room.

Valerie entered first and did not hesitate as she proceeded to the center. Although she did slow, and I felt her reluctance when she came upon the odd gathering. Jonathan was sitting with his back to us in a chair at one of the many long tables that inhabited the room. There was a large group of residents talking in low voices to the right of us and the table. There was a line of people stemming from where he sat. It looked comically like an assembly line or a long line at the bank or grocery store. As the occupants realized we were there, the voices and murmuring grew louder and more intense. Most were staring at me as if I were a T-bone steak, ready to pounce like starving dogs (or my heightened imagination coupled with my increasing insanity was getting the best of me). Some looked away guiltily, as if they had been talking about me before I came in.

Jonathan was slow to turn my way, and the look on his face stopped me cold. He was terrified. Of what, I wasn't sure. But his pale face, piercing stare, and expression of dread told me that there was much going on in this room and I was now going to be the center of attention. At once, before Jonathan could even open his mouth, a few members approached me quickly. A large man who looked like he had lifted weights and eaten hotdogs for a living was the first to speak.

"Well, I am not sure how Jonathan got you here but you came just in time." I didn't know how to respond and I was sure my expression showed how dumbfounded I felt. Jonathan hadn't told me to come; however, I got the feeling that he had told them I would be here. Searching for words backed with self-assured coolness I didn't feel, I stepped closer to the crowd and the looming man before me. Holding out my hand, which I wasn't sure he would take, I replaced my stupid face with one I hoped would calm and console.

"I'm sorry, I don't know your name, sir. I'm Adrianna, but I guess you know that." I sounded like an idiot but, to my immense relief, his face relaxed almost instantly. He shook my hand in return and stated that his name was Donovan Cleary. A woman with the reddest hair I had ever seen (obviously dyed) was next to shake my hand.

"We all know you, Adrianna. I am Lynn. Lynn Mathewson. I have been here quite some time, after being overcome by cancer. I left my four children and husband behind. If you know how I can keep tabs on them and not be sucked up into wherever the others went, then please do tell me. I told Jonathan the angel has given us the hope we all have been looking for, but he swears you are an angel too. So, tell us, Adrianna, how can you help us?"

I was completely gobsmacked now. She was strong but there was a sadness and desperation in her eyes that cut me to the core. I wanted to give her and everyone all the answers they needed. I desperately wanted to give them all hope and belief, and to remove the fear from their hearts.

They didn't deserve this. They had died. They had lost. Now they stood to lose again. I couldn't bear it.

I felt the energy in the room closing in on me, suffocating me. I felt bound, tied, and trapped. The room grew cloudy, although I could still hear the voices, shuffling of feet, and Jonathan's voice saying my name. My vision clouded further until I couldn't see anything at all. I felt weightless, my body lifting, floating, and flying up. I wondered if I was actually moving. Wow, would that shock everyone! They would really believe I was an angel then. (Oh my . . . I had finally lost my mind . . . and fainted too—I'm such a whack-job.) Maybe I was being sucked away like the others.

"Anna! Anna! . . .Look at me. Please, Adrianna." His voice seemed so far away, like there was a long tunnel between us, his voice echoing, hollow, and distant. I wished it were closer; I wanted to feel him near me. Surprisingly, I didn't feel afraid despite the fact I seemed to be flying away from everyone and I couldn't see crap. I felt pleasantly free somehow.

"Adrianna, stop! You must come back to us!" This voice was loud—too loud—and very near, as if I had been pulled back down by a string, tight and strong. My weight came back heavy and I felt a million pounds weighing me down. Something was gripping my arms, holding me, and as my vision cleared I could see Jonathan's astonished, worried face and the faces of the others standing over me. Crap, I had probably just gone into one of my can't-take-any-more-spaz-out fits—so much for the glory of the floating angel. Ha, dead people can faint (kind of). There was no end to the surprises.

"Are you okay? How do you feel?" His voice was clear now and comforting.

I found my words easily as if nothing had just happened.

"I feel fine, Jonathan, thank you." I wasn't sure how I felt really. I did feel fine—that is, I wasn't in pain, my weight had returned to normal, my vision was completely clear, and I was amazingly less stressed than I had been when I had first come into the room. Honestly, I had no idea what had just happened to me. As if invading my thoughts, Jonathan asked incredulously, "Do you have any idea what just happened to you?"

I couldn't stop the laugh that escaped my lips. The smile that plagued my face must have looked pretty out of place, considering the situation, but I just felt like smiling. "Ummm . . . I fainted, I guess.

Gasps and sounds of disbelief ran through the room as I looked around at their faces of uncertainty. I couldn't understand their reaction but it didn't bother me either. The feeling of contentment just seemed to increase with my smile.

"Holy shit, girl, you are an angel and you don't even know." I had no idea where the comment came from but it confused me. Had I done something crazy? Sprouted wings? Become earth-shatteringly beautiful? Would be cool if that had happened, but I sure didn't think it had.

"I don't really think I am much of an angel, everyone, but I am here to help. I don't have all the answers to what is changing this world, but I know that with Jonathan, Valerie, all of you, and my little friend Dierdre, we will find a way to stop what is going on and keep this reality safe. I am not super-special—guess I am kind of like Harry Potter, not powerful without my friends."

My speech was weak but I hoped it would be effective. The look of pure astonishment remained on everyone's faces, especially Jonathan's. I could not see Valerie but I was fairly sure she would look the same as the others, even though I didn't know why.

"Adrianna . . . you changed. I don't know what the hell just happened, but you changed right in front of all of us."

"Yes, I saw it; we all saw it. You lifted straight off the floor about a foot and you . . . holy shit, I can't believe I am saying this, but you had wings, a white dress, and, well, you glowed."

"She did glow, didn't she? And it was more like two feet, I think."

"Her hair glimmered and, wow, she was beautiful. She is still beautiful, but . . . ."

"Marge, shut up."

"She is amazingly beautiful and she is a bloody angel. I saw it before my very eyes."

"It has to be a trick, an illusion. Something they have set up to trick us."

The voices blended; after that, I couldn't make out any more words of angels or trickery. My head was spinning with what I had heard. I floated? Changed into an angel? What the hell was going on? Glowing? White dress? I looked down and was not overly surprised to see that I still wore my favourite blue jeans, with a hole in the left knee, and a black tank top that I thought made me look just a little more mature and sexy-like. Seeing no shine or light coming from my clothes or skin, I turned toward a very stunned-looking Jonathan.

"Jonathan? What is everyone saying? I don't understand all of this. What happened to me?" I felt my body waver as if an earthquake had just shaken under only me.

Without a word, he lifted me up into his strong arms and carried me away from the protesting voices. We didn't walk far, just to the other end of the long room that I had only been over to once or twice. There was a small easy chair in the corner next to a pretty, hand-carved pine table. The chair was soft and I immediately sunk in when he gently set me down. I closed my eyes for a second to concentrate on the feeling of sitting in a super-comfortable chair. Silly, but nice. I opened my eyes to feel the burn of emerging tears and to see Jonathan crouched on his knees in front of me. The expression he wore topped all: awe. I could see raw awe in his eyes that didn't belong in regards to me. It didn't upset me but I felt undeserving all the same.

"Adrianna, are you all right? How do you feel?"

"Ummm, I feel fine, I think. I just don't know." There was no desperation in my voice as I expected, but miraculously I didn't feel as confused or as shaken as I probably should have. (Yup. Confirmed. I am crazy.)

"Before we discuss you morphing in front of loads of people into a very beautiful, glowing angel, I have to tell you something."

I didn't have words. I just listened.

"Shit, Adrianna, I set you up, I guess. Before you came in here, I was battling with everyone and I told them that you would come and tell them how we would save them. I was desperate to say something, Anna; they were almost all against us. So, I told them I needed to speak to them one at a time, as you saw when you came in. I had almost gotten through everyone who stuck around. I needed to find out what they had been told. See, they have been spoken to by someone here who told them that there is an angel who has been wrongly banished from here, that this angel has promised them that if they give up their hold on this world and join him, he will show them the way to go back to the human world—not reincarnate, but go back. I know that this can't happen but they want to believe it so much. As you can imagine, many of them did not die a natural death."

He paused for a moment to gather his thoughts and allow what he had said to sink in. I was beginning to understand the atmosphere when I came in. Waiting patiently for him to finish, I relaxed further into the chair and glanced towards the main area of the room. Although we were partly around a corner, I could still see many people staring at us with the same look of shock and equal interest. They wanted me to come back so they could grill me. I guess I expected that and would want to do the same. If they thought they had actually seen me levitating and radiating an unnatural light, they must be full of questions, especially when Jonathan had them believing that I had more answers than I had. Oh my, I was in trouble.

"Ok, well, you had to say what you said; I know that, Jonathan, but what are they talking about? What did I do?" I hoped I didn't sound like I was passing him off because I really did understand why he needed to placate everyone and why he needed me to do it. I was not angry, though I was curious as to why he was so worried. I really just wanted him to tell me why they all went berserk and why he killed me, but that one would have to wait.

Jonathan began laughing quietly and very sweetly. I had almost forgotten how much I loved his laugh and the softness of his voice when humour took him over. "Adrianna, my dear, sweet, lovely, one-of-a-kind woman. You are an angel and you just proved that to everyone. I can tell by your reaction that you had no intention of lifting yourself off the floor, spreading your arms out to reveal the most silky, white, glistening wings while your whole figure glimmered in a light only you could radiate. Anna, somehow you became in form who you have always been in heart." He laughed again and his eyes were shining with tears. I was stunned silent. I could not speak. It was impossible to believe, of course, and I knew I would have to file this one away into the not-freaken-likely portion of my over-filled child brain.

"I don't know what to say, Jonathan. This is impossible. I am able to do a few unique things, that is for sure." (Ok, really, some of this crap was totally-wicked-cool.) "But that's absolutely insane. I floated? Had wings?"

Smiling, taking my chilly hand again, Jonathan leaned in closely towards my ear until I felt his warm breath travel down my neck.

"Adrianna, if I have one wish to fulfill, it is that you will eventually see yourself as I see you." His tone was musical, his words soft and quiet. It felt good for him to be so near, which made it an even bigger shock to my soul that the next words that came out of my stupid, ill-timed mouth were those that hurt us both.

"Why did you kill me?"

His breath caught in his throat. His body tensed and froze. I knew he had not expected this question from me right now as much as I hadn't expected to ask it. But it was done. The notion that Jonathan had ended my life was out there, on the table to be acknowledged despite the sensitivity. No matter the answer, I loved him. No matter the reason, he would always be one of the most important people in my existence. But I had to know. I had to. No matter how it might change the way things were, forever.

"Anna, I . . . ." His voice weak and shaky, he tried to speak to me. The thoughts and the feelings he had, the misery he was going through over a decision he had made so long before—I didn't know his exact thoughts but I did know him well enough to read the pain on his handsome face. The guilt I expected did not come. My need to know was greater than any remorse I might have felt for hurting him and putting him on the spot at such a bad time.

"Do you remember the first time you ran away from home? With your Strawberry Shortcake knapsack and your cut-off jeans. You must have been about six years old." He paused, his eyes a million eons away, and in another world completely. I nodded confirmation, although I don't think he saw me.

"I saw this in my dreams. I was about sixteen at the time. I had no idea why I would be dreaming about a little girl I had never met, from a place I had never been. It was one of the first dreams I had of you." He looked at me then, tears welling up again in his beautiful eyes.

"You were alive when you dreamed of me?" I asked incredulously. I wasn't expecting this beginning of an answer.

"Yes, Anna, I was alive and much different than I am now. I was selfish and reckless, but there you were: this pure, precious little angel that kept invading my dreams at night and thoughts in the day. I know now that the visions I had of you were real in your world and in your time. I did not know then. When my life changed and then I died, I still had thoughts of you; actually, you were one of the first people I thought of after I came here."

I couldn't find any words that would fit the scene he played for me in my mind. There was nothing to say. It was all so utterly amazing.

"When I first saw you after I had died, I didn't make the connection to the little girl of my dreams. Actually, it was quite some time before I figured that out. Once, when I visited you in the living world, I saw some pictures of you as a baby and a young girl in your mom's room. She had been looking through them. I realized then it was you who had been in my childhood dreams as well."

He paused again but I didn't interrupt. I was fascinated by his story.

"Sometimes, after I began visiting you, I made my presence known by leaving a flower or a note pad indented with a heart; I switched music on your stereo a few times, and even called you on the phone and hung up. You saw me as a bystander, as a faceless guy in the crowd, until that night at your school where I was watching you. I knew I would catch shit but I had to touch you and help you. It hurt, though. I have watched you for a very long time, Adrianna, wondering when you would come here and scared that I would be gone or have been here too long to be interesting to you when you did. Then you became talked about more and your importance was emphasized. I wasn't shocked, really; I knew you had to be special because of how you had enveloped my world in life and death. I fell in love with you—well, I think I always loved you, but then I saw you as my key to redemption and I prayed you would die."

He stopped speaking, ending in a cold, detached tone. He did not look at me but I saw the tears streaming down his warm cheeks. I saw the tight jaw, clenched teeth, and rigid form. He hated himself for what he was telling me and expected, maybe hoped, that I hated him too. I heard his every word, though I was still very confused as to why we were so connected and what that meant. Guilt rushed in fast and hard as the word "soul-mate" plagued my mind and heart. He did not give me time to finish allowing the overwhelming repentance to flow, as his next words ceased the torrent.

"I wanted you to die. I wanted to be the one who brought you here to the middle world to save us, although I did not know what you would be saving us from. I just knew that I messed up in my life, Anna. Badly. I was a monster. I also knew I was destined to do something important, something great to be known and remembered for, to erase the mess I made of my living life, to redeem me from being the murderer I am. I had to be the one to bring you here to save myself. It doesn't make much sense now but it did to me at the time."

The tears had stopped, in contrast to the growing pain. His apparent self-hatred did not allow the luxury of the release from continued emotion. He was killing himself inside.

"I saw you that day at the bus stop—even before you left home, when you put on that beautiful new shirt. I followed you down the road, wishing I could be the one to put the umbrella over your head if those grey clouds let loose the wet snow. I saw you smile at just about every person you passed on the road, despite being angry at your mother, despite the reasons you had not to be a happy, pleasant person. I wanted you to be here, and then you died. You just died right there."

I knew I should say something but my thoughts were confused and conflicting. He had wanted me to die but he didn't do anything to cause my death by his own hand—or was he not telling me something? Selfishness drove his interest in me: that hurt, I couldn't deny it. I so desperately desired the truth to be that it was not all his need to comfort himself here in his new world, or to validate his corrupt mind that caused me to be such an essential part of his existence. He abruptly resumed his explanation just when he thought he wouldn't speak again.

"Don't you see, Anna? I willed you dead . . . and you died."

Afraid that touching him would cause him further pain and torture, I kept my hands to myself but internalized my feelings. I wasn't sure if I even knew what I was doing, but somehow, I knew I had to: I began speaking to him through my mind. I didn't say 'Hey dude, get over it'; I didn't speak to him directly, but I more channelled my energy into thoughts for him. I sent him visions of our time here, my happy times, my best times; I sent him acceptance, affection, need, and desire. I knew I felt all those things and more for him, but much of my feeling had been stuffed down into a box with a tight lid since Valerie had told me he had taken my life from me. (Man-oh-man, did I have baggage.)

Now as I began to let go, many of those feelings came rushing back when I knew inside, deep in my heart, that although he had wished me dead, his hand had not brought me here. I knew that even if it had been his mind that had caused my demise, it was done because it was meant to be. I no longer saddened at the thought that I could have had more time in the living world when time there would not have been right. I was meant to die at that bus stop on that very day, no matter how it came to be.

Without my realizing it, those thoughts were being transferred to Jonathan as well. Realization of this came to me when his agonized eyes flicked to mine with lightning speed. His head was shaking lightly, and disbelief and disapproval were the driving forces. He was not ready to have forgiveness given to him. He wanted to hate himself.

I was about to reach for him again with my mind, and with my hand, when we were interrupted. A small man stood just a few feet away, wringing his hands nervously as if he had been sent to deliver horrible news. His expression turned to one of curiosity when he saw the look of torment on Jonathan's discomposed face.

"What do you want?" Jonathan snapped viciously when he saw his demeanour had given him away.

"I . . . I . . . ah . . . the others, they want to speak to Adrianna and to you. It is important that we finish discussing things with you and, well, after what happened . . . we have questions." The little man, clad in a blue terry-cloth jogging suit, with tousled, thin grey hair and rimless glasses looked almost comical being the one to speak for the group. I knew I had better handle things and give Jonathan a moment to gather his composure. He still looked ragged and drawn. I feared I had done more damage to him by asking him to explain than I was going through by not knowing.

"Arthur, right? I will come right away to speak to everyone. I am sorry for not coming sooner. Please ask them for just one more moment. Thank you very much for coming over, it must have taken courage." The little man brightened at being spoken to, although how I was being regarded was still very uncomfortable for me. I had to accept that I was looked upon differently than I saw myself deserving. Maybe I could empathize with

Jonathan more than I thought, but I still didn't understand the entirety of his self-loathing or what he could have done in his past to bring it on.

"Adrianna, I caused this. You need not face them if you don't want to." He spoke, but it was not him. He had changed in the moments we had sat here, away from the curious crowd so few feet away. Hardened. He was colder. It scared me.

"I will go, as I need to. I want to. They are our friends, our people; I will go and speak to them. Take a few moments, Jonathan, please, and think about what I said to you."

"You didn't say anything to me." He knew I was referring to the messages I gave to his heart and mind. I knew he had heard me: it was written on his face.

"Please, Jonathan."

"Yes, Anna. Be careful. They are not all with us. There is someone strongly against us. I might know who. I must take a few moments but I will be back very soon. Can you tell Valerie to join me by the well?"

Urgency plagued his voice as I saddened at his leaving. I wanted him to take some time but I didn't think he needed to leave the common room to do it. I was a bit scared but I knew it had to be done.

"OK, I will tell her." Wanting to say more didn't give us the time for me to do so. I wasn't sure that what I had to say would help, regardless. He had to go, talk to Valerie, clear his mind, and gain strength once again. Turning from him was hard but I didn't look back. I walked straight to the room full of people who had as many unanswered questions as I had, not knowing what to say but praying with all my heart I would find a way to persuade them that we would conquer our unseen enemy and prevail with our spirits intact. Along the way, I hoped I would convince myself as well.

* * * * *

It wasn't long before Valerie was beside him at the stone well in the courtyard. It was truly a beautiful place, with the green grassy areas separated by stone walkways leading to a massive old stone and mortar wishing well. There was no bucket hanging from a pulley, but the water was sparkling clear to the bottom. Flowers of all colors lined the edge of four of the main pathways. Green, trimmed berry bushes circled the yard in almost a perfect rim. The bright, warm sun added beauty and an essence of ancient times to this place, as if royalty had once walked these stones, weddings had taken place here, and creatures had discovered their souls in this very location.

She didn't speak as she approached. Jonathan didn't know if Anna had warned her of his mood or if she had sensed it, as she was so adept at doing. Leaning on the edge of the well, he saw her floating image appear next to his.

"Thank you for meeting with me. It must have been difficult for you to leave Adrianna there after all that has happened. It was difficult for me as well. I need to speak with you alone and safe."

"You know who is behind this, don't you?' As always, her insight astonished him. It would be important that she help him.

"We both know it is Lucifer himself who has created this whole mess. His hatred at being defeated and submitting to being in the mortal world for so long has made him even more bitter. We also know he has an accomplice, someone who has been helping him by delivering messages to the others and doing his work for him. I wasn't sure who until a few moments ago when I was, ahhh . . . speaking to Adrianna."

"Ok, we agree on everything so far. So, who is it?" Her impatience was clear and strong. She would not wait long for him to get to the point, which was more difficult after what had just transpired between him and Anna.

"You must listen, Valerie, and be patient. Adrianna and I were having a rather difficult discussion and, well, she was trying to . . . ah . . . make me feel better. She did something, I don't know how, and I don't think she knew either. She communicated her thoughts to me without speaking. Telepathy, or whatever you want to call it, like the elders. But in between the messages she was giving me I saw into her mind. I will not say all I saw, as it pains me to say that I have invaded her mind and she doesn't know, but I did gain a few pieces of information that I think we can use."

"Oh my, Jonathan, I did not know she could do that. Maybe she picked it up from Dierdre. What did you see?" Her questions were piercing but not as painful as the thoughts he had rushing through his head: Adrianna forgave him, even though she knew it might have been his mental demand that caused her death. She loved him truly, deeply, maturely, and without condition. She believed in him and was not concerned about his past. She believed they were soul mates despite his complete unworthiness. He also had learned that she feared love and distrusted relationships. She was afraid to be loved. It was too much.

"When Adrianna died, there was a woman, a large woman, at her death site, who looked at her spirit above the spot where her body lay and smiled at her. She is not of the living world, though like me she can venture there and be seen in the environment. She was also at Anna's funeral. I have not seen her here, so she is not one of us. She can only be of a different realm, a bad division of death. This woman that I saw in Anna's mind, I have seen before, though—in a picture drawn by Dierdre."

He stopped speaking to give Valerie a moment to absorb what he was saying to her, knowing she would be right along with him. This was confirmed by her nod and the squinting of her eyes. She had also seen the very drawing he had been referring to, a picture that seemed so insignificant that none of them had thought of it representing anything more than people little Dierdre had seen here.

It was a sketch of the large woman with curly hair speaking to Juniper, their friend. A friend they knew well, who was odd and a bit eccentric at times, but one they had trusted. A friend that Jonathan had spoken to that very day, just a short time before. She had claimed she was with them and would help them speak to the others to convince them that the dark angel was evil and would not make good on any promises. She had welcomed Adrianna and been an advisor to Jonathan in the past.

"So, Juniper is behind this with this woman. But why would she betray us, Jonathan? She lost her damn husband. What the hell is going on?"

Her surprise mimicked his feeling of disbelief. He was angry and confused as to her motives. Even if Juniper believed the dark angel, that he could send them back to the living world, she would not have her husband. He was a suicide, a lost soul. Now he was gone for eternity to an unknown abyss, unless she knew something that they did not.

"I don't know why Juniper has done this, Valerie, but I am sure that it is her. Maybe she has been told she will be reunited with her husband, or maybe she has just gone insane since he is gone for good; whatever the reason, she must be stopped. We must find a way to prevent her from speaking to anyone else or getting to Anna." He wasn't sure why a sudden feeling of dread came over him or why the hairs on his neck and arms stood up, but he did know he had a terrible feeling that something bad was going to happen or was already happening.

"You feel it too, don't you?" Valerie's question surprised him, as did her immediate intuition of his thoughts. She was good. Damn good. It always amazed him that a woman with her unique intuitiveness could be so darn snappy all the time. Didn't she predict that people would be angry at her brass? He nodded in affirmation and stared again at the well that granted no wishes.

"We must act quickly, but there are a few things I want to know first. Has Adrianna spoken to you about Willie, her mortal friend?" The question angered him before he could stop his jealousy over the very name from flaring up and becoming apparent. He knew he needed to stay focused, but it wasn't easy with her abrupt, random questions. He couldn't wait to hear what else she would spring on him before they went to find Anna.

"Not lately, no she hasn't. It isn't an easy conversation between us. Why?" Not sure he wanted the answer, he asked anyway, hoping she would move on to the next inquisition. He had given up on the idea of speaking to Valerie regarding the Willie situation, as Anna had not spoken of him in a while. There were times when he wondered if she was thinking about him. She would become very quiet and withdrawn, and there would be sadness in her lovely face. He never asked, for he knew it would hurt him to hear it.

"She has had another vision of him, which upset her drastically. She thinks of him often. I am certain her feelings for him are going to become a problem. I don't know how or when, but they will. You were selfish in not dealing with this earlier. You should have allowed her to talk about him and find a way to settle her mind. Just because she has issues that you do not understand or like to entertain doesn't mean you can ignore them. Love is not conditional." She stopped for a moment, judging Jonathan's anger level at what she had said. He remained silent though she knew he was fuming.

"Anyway, I wondered if her latest vision into her past was what brought out her courage in asking you if you killed her. I let it slip, Jonathan, and for that I am sorry. I was speaking with her regarding your feelings and it was brought up. I knew she was hurt but I settled her as best I could. Anna needs to concentrate and not allow her teenage emotions to intervene."

Valerie's arrogance was back in force. Jonathan's anger grew with her casual retelling of what had obviously upset Anna. At least he now understood where the conversation had come from, not that it made it any easier.

Struggling for days and days over his hatred for himself and his love for her had only worn him down. He was terrified he would lose her somehow, wondering if it would be for the best. She was an angel; she had proven that beyond any doubt, not that he had had any doubt. He had fallen in love with her, used her, possibly killed her, and now she had fallen in love with him. It was worse than a soap opera and way more excruciating to endure.

"Well, that's clarified now. Jesus, Valerie, do you ever stop and think before you speak? No, don't answer that. I am sure you don't. What other question do you have for me? And I hope it has nothing to do with my relationship with Adrianna." His anger subsiding slowly, he unclenched his teeth and allowed his face to soften, knowing that if Valerie were angered too they would get nowhere fast.

"I try not to think before I speak, young boy, or too many things do not get said. Your relationship with Adrianna is central, Jonathan, to what is going to happen. I have told you before that I have had visions of the end and a more recent one that terrifies me. I do not know if we all remain or how many of us do, but I have seen that Adrianna does not remain. I don't know if this is because she is wiped out of here forever by the dark angel or because she chooses to leave, but nonetheless she is gone. I also know that Dierdre will awaken, so we must see to her first." She must have realized that she was yelling because her tone reduced and she visibly calmed in front of him.

Stress, impatience, and lack of time were taking a hold of all their thoughts and emotions. Jonathan knew not to respond to the speech, as she was right. It wounded him deeply that she had seen this about Anna but he knew that not all her visions were absolute. He prayed, however, that if this one were right, Anna herself would choose her final fate and that he would be happy in her choice. He knew he would die anyway the day he lost her, by the dark angel's will or by his soul's own choosing not to continue without her.

"Let us go now. We will check in on Dierdre and then find Adrianna. When she is safe we will hunt for Juniper and the woman that leads her."

Without another word or shudder of fear, they set off to Valerie's room, where they both hoped to find the little sleeping angel awake and ready to help.

They were silent on the way down the halls, avoiding the common room. Jonathan's thoughts were centered on Anna and how they would find and defeat the dark angel. He knew that it was Anna who would defeat him but he had no idea how. She was not a warrior and had no skills for battle, though she was the strongest person he had ever met. If it came to it, he would stand in the way of any physical attacks but he was helpless in a battle of mental power.

Valerie's prediction ability may come in handy, and if Dierdre was awake, she may also be able to give Anna visions of a way to defeat their seemingly untouchable enemy, but they had to find him first. They had to draw him out by using Juniper and her unnamed partner. It was all so confusing and abstract to Jonathan. So much was not known and too much uncertain. They all assumed there even was a way to bring the dark angel to Anna so that she could attempt to overcome his eternal power.

Entering the room, their hopes were demolished as they found the little girl still lying statue-like on the sofa. She hadn't moved and she didn't still when they approached. Valerie took the spot that Adrianna had left, beside the child, and held her hand.

"Dierdre, it's Valerie. Jonathan is here as well. Can you hear me? Her voice was soft, gentle, and soothing. For a moment, nothing happened. All were still and motionless, as if the air had turned solid and frozen each of them in their spots. Slowly, Dierdre's eyes began to open. She blinked and immediately paled. Before either Valerie or Jonathan could react to this immense surprise, the little girl cried out.

Her scream was airy and weak but chilled them to the bone. Pure terror came from her body. Pure sorrow erupted into the air like a poisonous fog that hit them both at the same time. With the blast came a vision so clear it was as if it were happening right in front of them. They saw the dark angel, wings spread, floating above the ground just a few inches away from a figure crouched on the ground by his crossed bare feet. The jeans and tank top gave Anna away immediately and then they felt her fear.

The scene in front of them wavered with Dierdre's own fright for Anna, and Jonathan was almost sure it would dissolve before he could get an idea where they were. Just as he began to yell for the amazing child to hold on just a little more, she closed her eyes again, balled up her fists and sat strong on the edge of the sofa. Immediately the vision intensified, and not only could they see where they were but they could smell the grass and wet dirt, hear the rushing river water, feel the light wind as it whispered through the closely surrounding white birch and spruce trees; they tasted the power as it emanated from their rival, the horror from Adrianna covering their skin as sweat.

Looking closer, Jonathan could see that Adrianna was crying and she appeared to have a small amount of blood on her precious, beautiful mouth—the very mouth that never said a rude or unjust thing about anyone, the lips that sang a soothing tune in each word of encouragement, each command of love.

He felt rage, and pure violence surged through his very soul at the thought that anyone could raise a hand to such a genuine, angelic soul. He knew that he would stop at nothing to see her safe from the vile creature and safe from any evil that may try to warp her beautiful mind or damage her perfect form.

As suddenly as the wave had come, it receded, and they were gone. Jonathan stared at Dierdre without a word as the tears poured down her pallid face. It was all happening. All his questions on the 'where' were being answered, too fast—he didn't feel ready. But there was no "ready." You didn't prepare to face the darkest adversary of all history, the devil himself. There were no weapons, no magic spells, no unbeatable heroes to call upon to save the day.

There were just the four of them: a boy who had not yet learned what it took to be a man, with abilities only truly helpful in the living world; an old woman, bitter and cold but tuned in to the future; a child who could only communicate with a gift of sight, small, fragile, and afraid; and last, a young woman, kind, pure, and gentle, who knew no more dark misery than a life without a solid family and who gave relentlessly in hopes of making the world a happier place. The inexperienced quartet would face an ageless demon whose sole purpose was to bring evil to those around him.

Jonathan prayed that his legs would carry both him and the tiny child that he found himself scooping up into his arms, with Valerie's arms to guide. Delicate arms reached around his neck as soft brown silk touched his face and shoulder. The scent of strawberries and vanilla almost overtook him as he breathed in her gentle scent. Adrianna's sweet smell wisped off her as if they had been joined only seconds before, mixed with the sweet scent of innocence and purity from the little angel he carried.

As they zoomed out of the room and down the hall, Valerie did not ask where they were going or what they were going to do when they got there, for she knew he recognized the place by the river he had frequented to fish, to swim, and to dream: the waterfalls, where many had sat to drown their thoughts against the roar of the crystal-clear water, at this moment terrifying Anna. They would stand together. They would fight; they would do whatever they could, even at the risk of eternal death, to save Adrianna and Dierdre. They were the ones to move on or remain to move forward another day. They were special, and essential for life and the middle world to continue to exist. Jonathan knew this, Valerie knew this, Dierdre knew this, and somewhere deep inside, Adrianna knew this too.

12

Deception

"Oh, Adrianna, or should I call you 'Angel'? I am so glad you have come and have spoken with us. We knew you would, although there were some of us who doubted you. I never did, I swear." The plump woman sat beside me, her long brown cotton dress heaving with her raspy breathing. Beside her were about fifteen to twenty others who sat close by; the other middle-worlder's stood near enough to hear and speak.

"Oh, Mrs. B., please just call me Adrianna, and of course I am honoured and happy to come and speak with all of you. I only hoped to put your minds at ease and let you know I am here for all of you." I felt pleased and blessed that I was not only wanted in this large group of people, but needed as well.

The few hours I spent talking about my past, their pasts, and the future of our world was an experience that delighted me. We cried together, ranted together, and told stories of amazing courage, triumph, and retributions. Feeling like I had made a difference in many of the hearts of my fellow middle-worlder's, I was not taken aback in the least when Juniper asked to speak to me privately. I hadn't seen her for quite some time and, in the chaos, I hadn't really thought about her. Shamefully, I agreed to leave the others and walk with her to the outside gardens.

"You seem to have made quite an impact on them, Adrianna. May I ask what words of wisdom you gave?" she asked as we walked together out of the common room hallway.

An unexpected tone of patronization came with the words she spoke. I was unable to conceive of why she would be mocking my efforts to placate my friends. She of all people should understand their fear and sense of loss.

"Well, I didn't really say any one particular thing, Juniper, I just listened to their fears and thoughts, and then I shared my own. We all just talked and got to know one another." I had said more than that to some, but I felt a hesitance that I hadn't felt before when speaking with Juniper. Something was off. Bad vibes coursed through me as I tried to cease the feeling of danger. Juniper had always been a bit unsettling, but harmless. I had no idea why now I felt so different about her at this moment. The stress of the day's events was catching up with me.

We walked quietly after that, the tension mounting with each step. Although she seemed at ease, there was an edge to our stroll that I couldn't explain. On the occasion that she glanced at me and my slower-than-usual pace, I saw triumph and spite in her eyes. I was scared. Wishing I had never left the common room, I kept wondering if I should turn around before we reached whatever destination she had in mind.

"Have you ever been to the waterfall at the end of the evergreen field, Adrianna?" Her question was innocent and tone light and inquisitive. I began to wonder if I had been imagining my ill feelings.

"Yes, Juniper, I have. Jonathan and I had a picnic there." The memory made me smile inside. It was a nice memory. I kept my answers short, to not ramble in my nervousness, hoping she would continue to talk and give me more information regarding what this was all about.

"It is a beautiful place. I come here often to meet the angels. I hear you have finally noticed that you are an angel as well?" Again, her tone was light and friendly, although the statement almost knocked me off my feet. She must be referring to the experience I had in the common room, although I didn't know that she had been present for that. I felt a little embarrassed and at a loss for how to respond. Scrambling for words that would yet again explain that I did not feel worthy of the title 'angel' and that I really had no idea how I had managed to stun everyone by morphing into an angel-like being, left me quiet.

"That's all right, Anna—that's what your dear Jonathan calls you, isn't it? You don't have to tell me anything. I already know that you had no idea who or what you are. But I know."

Her disturbing words confirmed that she was not the same woman I had met before. There was hatred in her voice, a plan in her mind, and a dark shadow on her heart. Knowing I should not continue did not stop my walking feet from continuing the path to the waterfall. Feeling her free-flowing dislike for me did not push me to turn around and run.

Somehow, I knew that wherever I was being taken was somewhere I had to go. Whatever fate awaited me was unavoidable. I might elude my destiny but I could not hide from it forever, no matter how terrified I was. Strangely, though, I did not feel terrified. I felt almost relieved, as the pieces of one puzzle were coming together. Juniper was obviously the one. She had spoken to the middle worlder's; she had brought the edict and the false promises from the dark angel. She must have been given a promise too—a promise that she would be with her long-dead husband. An impossibility, but an irresistible dream that anyone would sell their soul for. The literature clearly stated that reviving the lost was not a power any could possess.

"So, you have been told you will be with Tim again." It wasn't a question. It was a statement. Using her husband's first name served its purpose. She stopped walking but did not turn my way. I could see her shoulders tighten and her hands ball into fists. Crap, it would suck to be punched by her. She was slim and rather beautiful, but had twenty pounds on me. She was furious.

"I will be with him again, Adrianna. It is inevitable. We are meant to be together and this time we have spent apart has only strengthened his love for me. He will be a dedicated, loving husband and he will no longer be lost."

She believed these words more than anything she had ever thought, felt, or said. Beyond doubt, there was no way to convince her that the dark angel had lied to her, and that of all the things he could do with his powers, reuniting Juniper and Tim was not one of them, even if he would want to do such a wonderful thing.

I was positive of one other thing, as well: I would follow her the last few yards to the edge of the waterfall, I would fall to the cool depths below if she pushed me, I would face whatever she had in store for me, but I could not convince her of anything other than what she believed, and I did not have the heart to try. Despite knowing I was powerless against whatever she had planned for me, I still felt relatively calm. I was worried, of course, about her flinging me over the edge of the waterfall, but I was sure death couldn't happen here. She was smug enough for me to be sure she had something in mind other than me going for an eternal swim.

The silence was rather annoying but I could see the end of our trek. Hoping that what awaited me would present itself as soon as we arrived made me quiver in an equal mixture of dread and anticipation. My thoughts suddenly shifted to Jonathan. I could feel his presence as if he were standing beside me, although he was nowhere to be seen. Maybe he was thinking of me too. The connection we had was stronger every day, even with the emotional turmoil we were both going through. I wondered what it would have been like if we had met while we were both alive (in my modern time, of course—I wasn't into the weird clothes from the '50s).

"We are alone for now, Adrianna. Maybe you should gather your thoughts." Her voice shocked me as it came directly from my right side. I hadn't even realized she was standing so close.

"What is this all about, Juniper? Why have you brought me here?" I asked the questions not really expecting an answer. She was always evasive, and surely this time would be no exception. My concentration and confidence had been shaken by my absorbing thoughts of Jonathan, his face still so fresh in my mind, his scent still filling my nose.

"Patience, dear one." As she spoke the words, she sat on the lone bench just feet from the edge of the roaring water. It was beautiful, though the noise, the smell, the feel of the spray and warm air did not completely register in my senses, did not divert my mind from what was happening. I could feel my arms growing heavy and my legs felt like immovable tree trunks. I wanted to sit, though not next to Juniper. I sat heavily on the ground a few feet away from the bench, putting distance between myself and the dangers just ahead.

My mind wandered back to Jonathan once again. I felt as if I were writing a point-by-point list for my parents of the pros and cons of our spending eternity together. One: I loved him; two: I believed he loved me; three: we had a spiritual connection unlike anything I had ever felt or read about; four: he was the only other boy my age here in the middle world that was utterly and devastatingly gorgeous, smart, and funny, and he smelled like every good thing in the universe. Now the cons: he might have killed me; he obviously did something rather horrible in his past, causing someone else's death and possibly his own . . . .

"He let his mother die, alone."

The jolt that hit my body felt as if a million needles pricked every square inch of my skin simultaneously—an all-over body tattoo. Pain, fear, misery, and a long history of hate besieged me at once. The voice was soft and velvety, old yet timeless. It could have come from a boy my age or my long-dead grandpa. It pierced my very soul with its familiarity and alien caress. I had been looking down at my crossed legs where I sat. I couldn't look up, though the pull of curiosity was fierce. Slowly, I lifted my head, though my eyes were closed. At once, I felt a heat on my cheeks and my hands began to sweat.

"Adrianna, you need not fear me, for I am not your enemy."

Again, I felt the pain of each syllable. He sure was my enemy, our nemesis. He would destroy us, this I knew to be a certainty. But he was arrogant in his belief that I would listen to him, hear his promises, and feel his control. I could use this to my advantage if I could clear my stupid head enough to look at him and then form a coherent word to speak.

"I would like to introduce myself to you, beautiful Anna." Hearing my childhood nickname come from anyone's mouth but my mother's or, more recently, Jonathan's, caused a violent shiver to course through me, not of pain but of wrongness, of disgust that he would refer to me by an intimate name. This anger began the process I had been waiting for. My mind began to clear of all other thoughts: Jonathan, Dierdre, the middle world, my past, and even my iffy future. All I saw in my head was the dark angel before me, though I had not yet opened my eyes. I saw him clear, beautiful, glorious, and menacing. I opened my eyes.

My vision came to life, a 3D version of what I had always seen. In form, he was tall, definitely over six feet. His raven hair was long and parted perfectly down the middle to flow silkily down to each round, muscular shoulder. To say he was handsome would be a sin. He was amazing—almost perfect but for the evil in his chocolate brown eyes. The California tan look worked well on him (not that I like that look) and gave him a god-like quality. His chest was void of hair, as were his legs, barely seen below his black robe. Why can't angels wear freaken blue jeans and button-up shirts? I laughed to myself at the thought of this being in front of me as a model for Tommy Hilfiger jeans.

He smiled when my own eyes met his. I smiled too. I didn't feel happy or friendly but I knew that I would have to show him a few courtesies to buy some time to decide how this was all going to play out.

I had no idea what he even wanted from me or if his main aim was to toy with and then kill me for good. Regardless of how this would end, I meant to prolong it until . . . until what, I didn't know.

"You are even more striking in death than in life, Anna." I forced the cringing feeling to not reach my still-smiling face so as to keep things as pleasant as possible (yeah—crazy, I know). He was completely fooled by my continuing smile, although I sure had no formal practise at this or any other type of pure deception.

"You are Lucifer, I suppose?" The double gasp I heard nearby alerted me to the two women now standing a few feet to the left of the dark angel. Juniper stood with the "O" look on her face next to a large woman that I, unbelievably, immediately recognized as the woman from my death scene, the one who had smiled at me as I was suspended over my bus stop. She was also the woman I had seen at my funeral and had had daydream glimpses of since I died. She was the middle lady between Juniper and the dark angel. She was another dark angel, though clearly not as powerful or important. She was not smiling at me now. Her face held an arrogant disgust that made her double chin look almost cartoonish in appearance.

"You do not speak to him," she growled.

"No, Gloria, it is ok for her to speak to me, or any of us. Adrianna is unique and deserves our respect." The looks of horror at my directness turned to expressions of complete fury and disbelief. They had obviously never seen him being kind, or showing consideration for another, especially a simple teenaged girl who had no idea how to speak or behave in the presence of something too terrible for the imagination.

"Please excuse these lovely ladies, as they seem to perceive you as some sort of enemy. However, I do not share their opinion, nor do I believe you wish us any harm."

This was a turn of events I wasn't expecting, though I did not know why I should feel surprised that he was trying to manipulate me, as I was attempting the same with him. He was not acting the way I expected, though; with the degree of damage he could obviously do, it seemed irrational that he should take the potential victim's fearful role just to make me feel powerful and special. It was confusing. However, this situation also indicated that he wasn't reading my thoughts, as he did when he first arrived. (Score for the good angel team.)

"Juniper is my friend, or at least I thought so. Gloria and I have never met, though I have seen her. I am happy to meet you, Lucifer, although I am not sure what you want with me." Again, my boldness shocked me as did my solid composure. Holy crap, where did it come from? I was like a different person inside the same body. I felt like the same me, but the words that came out of my mouth and the stability that I seemed to possess were very new to me and seemed to please the dark angel while angering his two companions equally.

His laugh was even more fluid than his speech. Sounding like a song, his light chuckle continued long, ending in a wide smile revealing, of course, perfect teeth.

"You are a direct young woman, aren't you? Well, I am pleased to see that you show no trepidation towards me, as most do. I have never thought of myself as an intimidating or imposing figure, although I have been told that I am." His smile took on a hint of sarcasm, ringed with a huge ego. He loved himself for it. Oh, how I detested self-centered men.

"What I want from you? Not from you, dear girl. I just want to know you. To meet you in person and watch you uncover your potential. To see you flow from the delightful young woman you are now to the powerful, strong, unequivocal spirit you are destined to become."

"Lucifer, I honestly do not believe a word you are speaking to me, nor the false innocence behind them. I believe that you have sent Gloria here to give Juniper the hope that she will be reunited with her husband if she helps you make the other middle-worlder's believe other such promises, so that they will leave this world that you created but no longer want to exist."

Whoa, the speech was out and presented before I could even grasp what I was saying aloud. On occasion, I would think to myself, usually after becoming angry at someone rude, many fancy words that would all string together in a perfect sentence to shock everyone speechless . . . but I never usually said them aloud. I was usually non-confrontational, a passive person who didn't much like to rock the boat, but . . . wow, I was on a roll—a very terrifying, un-chartered roll, but a roll just the same.

The smile stayed on his lips but his eyes darkened further (if that was even possible). I had angered him, I was sure, but I was also sure I didn't care. He had the ability to wipe me out regardless of what I said at any moment. I had no patience to play sweet, innocent, I-know-nothing little Adrianna. I was feeling strong despite the nausea rising in my always-empty gut.

"My, you are a confident little lady, aren't you? I am astonished you have been fed such awful stories about me, my dear girl. The things you have been led to believe are simply not so. I do not wish this world not to exist; on the contrary, I sent someone here for this world to always exist as I have created it. However, things have changed here, much to my dismay, and I felt a need to come and . . . let's say . . . sort it all out."

I was confused by his words. Sent someone here? Who? Was it me? Did he send me here? I didn't think that could be it. Maybe I was wrong, and more confused than I thought. Maybe Jonathan and Valerie were wrong about Lucifer, that he wanted to destroy this world. He seemed to be telling the truth. But what did I know? I couldn't be counted on to properly read the darkest angel of all time.

"Who have you sent here? What needed to be sorted out? You took people. You have already destroyed souls, old souls—how can I trust anything you say?" I was becoming unhinged and knew I needed to get a grip. There was no point in asking these questions without being sure I would be able to decipher the truths. I wanted to rein in my enquiries but didn't, at the risk of looking like the idiot I felt like.

"Anna, Anna, please. We have only just met. We have much time for questions, as I may have some for you as well." The gleam was back in his eyes, as he knew he had rattled me quickly from my stoic composure. I had to gain back some ground, though I didn't know how. His first question for me dissolved what little hold I had left.

"Do you not want me to finish my beginning statement, when I arrived, about how Jonathan aided in the death of his poor, lovely mother?"

He had taken a few steps closer to me so that he was standing just over top of me, looking down. His slightly extended, previously hidden black wings were blocking out most of the sun so that I was bathed in his shadows. I saw him point without a glance towards the bench by the falls for the two women to take a seat there. Their incredulous expressions at our open communication had not once left their faces. I almost laughed aloud at their equal appearance, frozen in place. But I couldn't laugh, not with the curiosity burning in my chest. I had wanted to know about Jonathan's mother ever since I met him. I also wanted to know what he had done in his living past that had caused him so much agony. Now I had the chance to find out. Lucifer continued without a response from me, not that I had one to give.

"His family was quite poor, as I am sure you know. His father had left, his brother was dead, his little sister and baby brother were looking for help and his ill mother had a tremendous burden on her tiny shoulders. Not a vivacious woman, and not healthy, she frequently relied on Jonathan to aid in the care of his siblings, the care of the home, and eventually bringing the money needed to feed them all."

I would not have asked him to tell the tale, as I felt it was not his place, but as the words flowed from him I was captured. I knew most of what he was saying, but I was certain he knew more than I. Would it be wrong of me to hear the tale that Jonathan himself could not bring himself to reveal to me? It didn't matter how intense my guilt grew, as I would hear this story to the end, and I knew he would revel in the telling.

"Jonathan was a very typical boy in his teens. He loved the dangerous side of life and coursed through without stopping to see what he was leaving behind, though I must give him credit where it is due, as he did take wonderful care of his little sister and brother, and did the best he could for his mother while he was home. But it grew too much for him, the poor boy, and he found solitude and happiness outside his responsibilities. His mother was the one to suffer. And she did suffer. He left her alone, ill, poorly nourished, and without care, for he was absorbed in his own selfish ideals of life." He paused, allowing his tale to take the effect he knew it would. Seeing the look of curiosity on my face, he continued, sharing information he shouldn't know.

"Jonathan involved himself in breaking the law, using young women to pleasure himself and living only for his wants and needs. All the while, his mother grew weaker and weaker. When the disease she was consumed with took her life, she was alone at home, awaiting Jonathan's return with her medication. He didn't return until early the next morning, long after she had expired. Because of his reckless lifestyle, he was unable to take legal care of his siblings and they were lost to him. It is a very sad story, to be sure. Now do you see why I am saddened to hear of the information you have been given, knowing where it has come from?"

I felt again the stabbing sensation I had had when he first spoke those dark words. Tearing into my heart, I crumpled into myself, wanting to shut him out and close off the doorway to my heart he had so eagerly pushed through to cause such raw sadness. The words were true, this I knew. The glory he felt in delivering them to me was as evil as he himself. It made him happy to hurt Jonathan through me. My pain equalled his triumph, and Jonathan's story, his loss, was a joke to him, which he passed off with a façade of understanding and false sympathy. I could not sit here and allow him to continue building his torrential ego at the expense of the one man I loved and needed more than ever.

Reading my face, he grimaced at my internal defiance. I had no preconceived notions that he could or would be able to exactly read me, like Valerie or Dierdre, but his intelligence surpassed mine by a long shot and I could often be read like an open book. However, it did seem coincidental that twice since he had arrived he had instinctively known that I had been thinking about Jonathan. There had to be a connection there and he was aware of it. I had to be careful.

"So, you see, my lovely girl, he has a sordid past, filled with self-inflicted, painful memories and mental anguish from the time past. It has seeped into his soul and been passed through his words to you. I shudder to think of how warped your mind has become because of this unjust interference." His broad chin loomed over top of me when I lifted my face to glare once again at his beautifully carved tyrant's face. His smile was as evident as his flowing black hair. Courage bubbled inside me as if it was being pumped into my very veins. I felt stronger with my growing anger, despite the continuing heaviness in my limbs.

The thick feeling in my legs worsened and I suddenly felt myself sliding slowly back down to the ground. I was not sure if I was about to have one of my stupid and usually untimely fainting attacks, but this felt different. I felt my knees hit the ground and my body follow.

Despite knowing I would look comically hideous to the observers, I pushed myself to get on all fours in an attempt to stand and face him (or as close as I could, seeing that he would still tower over me in height).

I heard one of the two women stifle a laugh at my feeble attempt and it dawned on me that he was causing this. He had me lowered by his power. Invisible restraints bound me, cowed me, forced me into physical submission in hopes my mind would follow. Now that I knew this, I would definitely do something about it.

A look remotely close to shock filtered through his smug face as I slowly but surely rose to my feet in front of him. Locking my gaze with his, I thought not of Jonathan, not of how weak my legs felt, not of how ridiculous I must look facing the most powerful creature to roam this world, but of how good it felt to stand and be strong in my own right.

Overcoming his shock at my ability to defeat his mental bondage, he simply smiled once again at me, like a father to a daughter that was learning how to walk at a young age.

"And the mysterious Adrianna rises as the angel she is before me."

His voice was still fluid but held an edge of contempt. I could not avoid his stare or the growing fear that I was pushing limits he had not knowingly set. I stood, meeting his gaze with a fading confidence and a prayer that something would happen to break his hold over me.

And something did. I almost immediately felt the release of my bonds. I was clear of his power once again. Whether he allowed the hold to dissolve or I could conquer it, I wasn't sure, but I was free.

"You may tie me up and even stuff a rag in my mouth to keep me from speaking, but you will not ever control me. I am free, as is each one of us middle-worlder's, even if you take us all away from here and send us wherever you sent the others. We do not belong to you—our souls are ours."

My poise returned, and I could say with conviction how I felt and to make him hear the truth: that he had not won and would never win.

"My effervescent little Anna. I created this world. Without it you would all be dead and gone or have moved on to the next pathetic existence in the living world. You are nothing without me. I have allowed this world to continue, securing my place here with the help of my friends." He subtly indicated Juniper and Gloria, who were still sitting silently on the bench. "And, of course, sweet little Dierdre."

He smiled that familiar, wicked smile, full of self-righteous power and godliness. He knew he had the upper hand. It infuriated me that he would have any connection at all to Dierdre but it did not surprise me either.

"You see, Adrianna, after I was so rudely banished I spent many lifetimes as a living soul. I gained knowledge on how to defeat and control those pitiable mortals; I also found them to be quite predictable and usable. I could not lose my connection to this world that I had created, so I planted my seed here. My own child would guarantee that I could come here any time I wished to change what I felt needed changing."

Reading the mixture of confusion and horror on my face, he softened his words and slowed his speech. I wasn't having trouble hearing or understanding what he was saying but my heart did not want to believe his words. His child? Dierdre? He had sent her here, when she should have been able to move on and return as another innocent living being. He was purely despicable. But how was she his child? It seemed impossible. Seeming to realize my struggle to believe, he continued in explanation.

"You see, Adrianna. I was many humans. I have led many lives. I have been a father, a mother, a teacher, a saviour, and a murderer. I had to step in as little Dierdre's father in her life. She was needed here to secure my place. So, she is here and will remain here. I have been satisfied with the stagnancy of this world until recently—actually, just before your Jonathan decided to bring you here." He paused, gathering his emotions and thoughts. Looking angry but composed, he drew a deep, useless breath.

"It seems that there can only be a certain number of souls here, and that quota has been filled. A few souls I wanted here never arrived, so I needed to remove some to make room, that's all."

He paused again, looking innocent as a child that stepped on an anthill just because. He continued, his voice slick with sweetness.

"I had no idea you existed until Gloria came to enlighten me. But you were harmless in the human world. Now you are here and you are disrupting the souls, bringing freedom with you. That does not belong here. I decide who stays and who goes. I allow some to remain, some to move on, and some to cease to exist when they perish. Jonathan made the mistake of deciding for you to come here. Now I must dissolve all of you and choose to start over again." It was very apparent that his anger was growing: his wings, which had stayed mostly hidden behind his robes, were now almost fully extended; his brown eyes were now as jet black as his hair; and his hands, once relaxed, were now balled into fists.

I was beginning to see why he was here and why he was taking back the control he had been losing. It was a power struggle, and he would not lose again. I also knew now he meant to remove us all. Every soul I had reached and spoken to. Each person in that common room would cease to exist if I did not think of a way to stop him now. But what the hell could I do? A stupid teenage girl with a few interesting parlour tricks and a heart full of foolish hope, love, and dreams. I had no weapons. Or did I?

"So, you mean to destroy us all." It was a statement and he remained silent.

"And you used a poor, innocent child to allow you entrance into a world you created so long ago because you wanted to add certain people here and you didn't like my interference when I arrived. Is this correct?"

I was starting to form an idea in my head but needed much more information before I could run with it. Even if he confirmed my thoughts, I had no idea how I was going to proceed.

"I have just said as much, Anna. I need not repeat myself to a woman of obvious intelligence." He was back on the charm trick again, having overcome most of his anger with his arrogance. He thought himself to be unbeatable despite being beaten once before, and probably couldn't conceive of being challenged or conquered by a girl. Typical man.

"Well, I am obviously terrified, as you are the most powerful angel of all time. I just want to understand what you are saying to me before you delete me from eternity."

"I have not decided to delete you, sweet Anna. Although you are more misdirected and confident than I expected, I see great things for you—for us, upon our uniting together. We can own this world and its new inhabitants." Though I could not see the response from the ladies on the bench, I heard their disapproving gasps once again. The dark angel turned impatiently to confront them.

"I see no reason for either of you to be sitting there. Go and find our other friends and my daughter. You both will be granted your wishes the end." With a fast wave of his hand, they were gone and his attention was back on me. I felt more afraid somehow with them being gone, not that they would have helped me in the least had they stayed, but their presence had given me comfort all the same.

"Why can't everyone stay if you wish to regain control of their souls and rule this world with me?" It was an honest question, despite the obvious fact that I would rather die a million deaths than spend even a minute ruling with him over anyone or anything.

"They have been tainted; some are too gifted without a cause, many have been here long enough and I feel it best to start fresh"

"And what of your daughter?" It was nearly impossible for me to ask the questions calmly with my insides ripping apart at the very possibility of life without Jonathan and Dierdre. "And Jonathan as well?"

His laugh came short and unkind, though a simple smile followed it.

"My daughter will stay forever, as she is our key, Anna. And Jonathan . . . Anna, did you know that after he allowed his mother to die, she ended up here?" He must have seen the look of complete shock on my face, as I couldn't believe Jonathan had not told me his mother was here.

"She was here, in the room for the lost. She was taken many years ago, during a time when one random person would be taken over a period of time to keep the numbers down. Jonathan hadn't been here long when she left. He didn't even know she was gone."

"But she didn't commit suicide, so why was she in that room?" I asked, thinking of the poor woman and of my brother.

"You see, Adrianna, that room is not only for those who have taken their own lives. It is also for those souls who are lost from extreme grief or remorse, or for those souls who are pushed into this new existence, as you were by Jonathan. Sometimes, they go insane. Jonathan's mother was . . . well, she was lost in the few days before her death. Grieving over her son, husband, and being abandoned by Jonathan. I believe she went absolutely mad. Her heart broke."

There was mockery in his voice and irritation at my concern for Jonathan. I was starting to see that his occasional softness and kinder words were not all a manipulation. He did truly believe that I would join him and we would rule together. It was almost inconceivable that he could be so delusional.

I had a choice here. A very terrifying choice. I could play this game and pretend that I was actually considering his proposal, or I could stand my ground (until he crushed me) and tell him to go back to hell or wherever he had been banished to before. I didn't know what to do. I wanted to talk to Jonathan and have his soothing arms comfort me. But I knew I must not think of him: any plan I had would be void if the dark angel knew my mind had strayed to Jonathan once again.

Turning my thoughts toward Dierdre, I tried to concentrate on where she was and if she was ok. I had felt earlier that she was awake but I had lost the connection. It was very difficult to concentrate with him standing here watching me.

"You have made friends here and will be sad once they are gone." The fake compassion in his voice was almost convincing and I felt tears form in my eyes.

"I have been too harsh, too power hungry, Anna; please allow me to apologize to you. I am not the evil monster I have been portrayed as. I have said things now, here, to you, that I shouldn't have said. Would you believe I am just angry and sad at not being the one to meet you first? Not being able to have what I fought so hard for, for so long?" His questions were rhetorical, I thought, and I didn't respond. He was changing his tactics, seeing that he was getting nowhere. But why would an angel of that much magnitude even consider a switch in power?

"I was once the most respected angel. Then I became greedy. I am not insatiable anymore. I just want what was created by me. A place for those souls who have been chosen to die to be able to watch over their loved ones until they arrive or go on to exist once again—not so unlike you watching over your loved one, Willie." At the mention of Willie's name, the tears I had been harbouring sprung. I was unable to hold back the quiet sobs from escaping my body. I was dying for all those souls who were already lost, those who were left to be ripped from their after-world, for Dierdre, for Jonathan, my friends, and for Willie.

"I sometimes become lost in my impatience to have it all back again. And now to share it with you, the only female angel with powers not unlike my own but with a heart that I lost many eons ago and am barely just finding again."

His words were impossible to believe but somehow were breaking into my shell. Maybe he was telling the truth. He had been a great angel. He had created this beautiful place and allowed the occupants to add to it from their experiences and memories. There must be a semblance of his heart left; he could not be just pure evil. Could he?

"I am not evil, just in need of direction and understanding, Adrianna. Please believe me." With this last declaration, he wrapped his strong, warm arms around my cold shoulders. I could feel his chin rest gently on the top of my fluffy head as he nestled my head onto his chest. He smelled vaguely of ash mixed with a floral bouquet. It was a confusing odour, heightened by an even more startling feeling of comfort. To be held by Lucifer, the dark angel, the devil, was not at all what I expected. It was still frightening, but pleasing as well. The confusion I felt a few moments before was tripled in intensity now.

But I needed to think. I needed to be alone, or at least not in his presence. How could I escape him?

"Would you like time to rest, Adrianna?" His abrupt, intuitive question caught me completely off guard. I was sure he could not read my mind, as many of my earlier thoughts probably would have pushed him over the edge, but he sure had a good grasp of my emotions and what I could be feeling.

"Ah . . . yes, Lucifer, I would." I had nothing else to say to him. I was too confused for words to explain. His face surprisingly showed understanding.

"I will leave, as I have a few things to attend to; you stay here and allow the roar of the water, the beauty of the birds and flowers, the soft warm fragrant breeze to relax you into a restful state. I will have no one bother you." With these last words, he was gone, though seconds before his sudden departure I could swear he had become almost transparent.

He was to be believed about one thing for sure: once I dried the remaining tears on my cheeks and allowed the atmosphere to absorb me, it did calm me. I felt tired, like I had just run twenty miles without stopping (I could never actually do that), and strangely I had a bit of a headache but I was slowly relaxing as if he had slipped a tranquilizer into the air. Each breath rendered me less anxious and more serene. After an undetermined amount of time, I no longer noticed anything around me and was completely at peace.

13

Lucifer and Juniper

"Take the ones I have picked, who will be useful to me in the living world, and move them forward; start the process of eliminating the ones who are truly against me, and keep a few who have stayed true to me." He spoke harshly and quickly as if he was running out of time.

"You must take them one at a time, and slowly, as my power is strained." Both Gloria and Juniper nodded to his order with questions left unasked in their minds. Juniper was the first to speak to him.

"What of Adrianna, and Dierdre, Jonathan, and Valerie?" She asked with fear of his impatience, looking around to make sure they were unobserved. The path to the common room was empty; most were still gathered inside, still talking about the amazingly beautiful, white angel that had formed in front of their very eyes just a short while earlier. They were mystified, and most of them were completely under Adrianna's spell of hope and goodness. It was seriously pathetic and rather disconcerting to Juniper. She was too afraid to tell the dark angel that he had but few followers left, that Adrianna's speech and one-on-one story swapping had changed their fears and soothed their minds. He would be furious to know that even some of those he had chosen to move on to aid him later would be no good to him now. But she would not tell him this. She would quietly do his bidding and hope that the process would be quick.

Looking at his transparent form was unsettling as well, though she knew he was only able to be here for short periods of time as his form was still bound to the living world. His banishment so long ago had bound him to the mortal world unless Adrianna could be convinced to agree to rule with him; then he would be free to fully monopolize this world as well as many other states of reality. It was frightening to think this deceptive, evil immortal could have so much power over so many living and non-living souls. But what did she care? She had been guaranteed a long life with her beloved husband and to be reunited with him in each life after. That was all she wanted, all she ever asked for; the rest was up to him.

"I will deal with them after my absence. However, you are to keep them apart. They must not reach Adrianna, and she will not go to them—of that I have made sure. I will return soon and finalize the eliminations."

He was gone as quickly as he had arrived, leaving nothing but his orders behind. The two women stared at each other, both wishing they did not have the job at hand but both knowing their existence depended on it.

"I will start in the common room, Juniper." Gloria spoke first, quickly, and quietly. The large woman had been helping Lucifer for much longer than Juniper, since she had been killed trying to kidnap the third child that she had felt entitled to. She was not unlike Juniper in the sense that losing her own children in an accident had caused her so much pain that she would do anything to have them back, although Juniper could not see herself kidnapping others' children to fill that void. Lucifer had found her just after she was gunned down by police, and recruited her to help him keep an eye on the middle world when he could not. She had taken care of the first few sets of eliminations, prevented many from moving on, and kept order until Jonathan had interfered.

"You must find Jonathan, Valerie, and Dierdre and keep them from finding Adrianna. I am doubtful that Valerie will be able to "see" Adrianna now, but to be sure, keep them apart. They do not know you have any part in this, so befriend them again and keep them distracted." Without looking back, she opened the outside entrance to the common room and closed the door behind her.

* * * * *

"Jonathan, I see her up ahead; she is sitting on the bench by the falls. Couldn't miss her hair." She laughed excitedly, like a child who had been lost finding her parents in a department store. Valerie walked faster, almost breaking into a run. Struggling to keep up with her, holding on to Dierdre with all his might, Jonathan kept his eyes focused on Anna, sitting alone on the bench. He wasn't sure why she had not noticed them coming, as they were now in plain sight. She seemed frozen, like a delicately carved ice sculpture. She was truly radiantly beautiful, even when she was sprouting magical wings, with sun filtering through her skin. If she only knew how she was perceived, who she really was.

Jonathan couldn't wait to touch her, hold her, kiss her luscious lips.

They reached her as Dierdre opened her eyes to stare at the stone figure. She frowned slightly and searched Jonathan's face for an explanation. He had none to give.

"Anna?" His voice was loud but soft, as if speaking harshly would shatter her very form. She did not turn. She did not even flicker. Her eyes were closed but a faint smile played lightly at the corner of her parted lips. She appeared to be in a trance, though he was relieved to see she was not in distress.

"Anna? Can you hear me?" he asked again, not really expecting an answer but hoping beyond hope that she would not slip into a deeper state, as Dierdre had. He needed her; they all needed her. Sitting on the bench beside her, he reached his arm around her shoulders and drew her to his chest. She leaned with no resistance and he was almost sure she even snuggled in a little, but that could have been wishful thinking. The feel of her body sent shivers through him and a happiness that was enticing, unexplained, and desired more strongly than a drug addict's cocaine.

"Adrianna, we are here for you—Valerie, Dierdre, and me. No one else is here and you are safe." He wasn't sure what to say to make her feel secure enough to break out of whatever trance she was in. Words, touch, love, he would give it all to her. Anything.

When she didn't stir, or react in any way to the spoken words, Jonathan sighed with fear that she would not awaken before they were found, as he was sure they were being sought. The feeling had been growing, since even before they had arrived to find her, that they did not have much time left.

As if sensing his anguish and impatience, Dierdre surprised them all by jumping down on little legs that hadn't touched the ground since she had succumbed to her own coma. As steady as she always was, her little form was beside Anna as if she had always been there.

Dierdre took Adrianna's free, limp hand and held it in her own. What happened next was not expected, but then again, they had come to realize that with the little angel's influence, anything could happen. Jonathan felt the current through Adrianna's still body as soon as Dierdre's tiny frame shook with it. Jonathan was astonished to see a faded image begin in his mind. It was not clear but it was there as if he were conjuring it himself. Anna's body stiffened within his arms and he knew she was experiencing it as well. Hoping that it would bring her around, he concentrated on the flitting scenes in his own head.

The room was dark and dingy but had the potential to be bright and cheerful, walls painted as yellow as the sunshine, a shag carpet over old linoleum, a surprising white. Cluttering the corners were dozens of toys that matched the beloved A.A. Milne characters on the walls. A wooden crib stood against the west wall, between two large curtained windows that would reveal a beautiful setting of the sun if not covered tight by heavy drapes. The rocking chair in the corner looked as if it had seen many years of use as well as a few teething puppies. Although the room was favourable for a child, even in the gloom, it was sinister and foreboding.

Sobbing loudly, a red-faced little boy in a blue one-piece jumper stood shakily on his tiny legs, holding on to the top rail for support. The tearless sobs turned into a wail, and in the distance loud footsteps could be heard. In a split second, a tall man entered the room. The first thing Jonathan noticed about the man was that his features were dull. Neither his face nor hair could be distinguished. The faceless man approached the crib with a slow stride, his hands balled into fists of rage. Yanking the screaming infant from the crib, he began to shake him with a horrific force.

Jonathan could feel Adrianna's body begin to shake with rage and despair, feelings he himself was enduring with her. The expression on his face must have given away two things: that he was definitely experiencing the vision that Dierdre was projecting to Anna, and that this vision was so repulsive and disturbing that he wished beyond any wish not to have to witness it.

When what was happening became apparent to Valerie, she went closer to the traumatized trio on the bench. Having no room to sit, she stood behind Dierdre and placed her hand on her fragile shoulder. The image hit her with sheer force, almost knocking her to her knees. Holding her ground, she met the stare of Jonathan and knew he understood what was happening to her as well. The vision continued strong, even with the additional observer.

As the vicious shaking continued, the screams stopped. The child's little body became limp and floppy in the man's hands, though he did not stop. Whatever possessed the tall, faceless man did not cease with the death of the baby he held. Hours seemed to pass before the shaking became slow and finally stopped altogether. The man stood staring at the inert infant still in his grip before dropping him in the crib.

No expression could be seen on the killer's blank face as he turned to see the little brown-haired girl standing in the doorway. The look on her face, however, said she had been there long enough to witness the cruel, brutal death of her baby brother.

Whether it was the recognition of the little girl by Jonathan and Valerie or the scream that took place first was impossible to determine, but Adrianna's gut-wrenching cry was heard loud and clear. Surprisingly, though, the vision did not end. Anna, now awake and coherent, stared at Jonathan as if she had not known he was there. She hadn't known. She hadn't known any of them were so close, touching her, trying to comfort her. Before the appalling scene had plagued her mind there had been nothing. Emptiness—not sad, just barren. It was all clear and real now. Jonathan was cradling her in his reliable, loving arms, little Dierdre was sitting to her right, holding her cold hand and looking even more alive and precious than ever before, and Valerie stood, as horror-stricken as the rest of them, behind Dierdre, holding on to the bench for dear life.

Dierdre knew they had seen her in the image. Tears slowly escaped her mystical green eyes, as she knew there was more yet to see. The pause in the movie before them was short, and it resumed slowly as Dierdre-in-the-doorway began to cry too. The man did not yell at her, as was expected, and he did not immediately advance on her; he just stood as blank as before. Jonathan worried in incoherent thought that the inability to see his face meant something. Meant something bad.

As suddenly as the vision had commenced, it shifted. There was now a hotel room, fading and blurry. The walls were brown, as was the bed. Twenty bucks was the most one would pay to stay in a dive like this. The figure on the bed was hunched over what looked like a writing pad, but it was unclear. The person was male, dressed in black, and with dark hair. As the vision cleared ever so slightly, it was apparent his hair hadn't seen shampoo in quite some time, nor had his clothes seen a wash. On the bed lay a half-full bottle of what was most likely whiskey, a small pipe, a long rubber band, and a syringe. He was writing, feverishly, as the pen flew in nearly illegible scrawl across the lined writing tablet.

The only words that could be deciphered before the scene changed once again was one line of revulsion to one person . . . 'I will see you soon, Adrianna.'

As the image faded, Adrianna's whimpers could be heard, almost as loud as her moans. Jonathan's embrace could not comfort her against the mental anguish being thrown at her in lightning-fast pitches. Her gentle heart could not grasp the maturity and acceptance it took for Dierdre to replay the most agonizingly painful memory so calmly almost anyone could have to bear. The little girl's face was pained but composed, her tears still cascading down her pallid cheeks.

When Willie's figure interrupted the previous hell they were seeing, it was like the last fibre of her sanity was being strung across miles, vibrating, tearing, and threatening to snap. She felt rather than saw the look shared between Jonathan and Valerie, and she knew that the image of Willie was not hers alone. She also knew what they must be thinking about the objects on the bed: alcohol, drugs, death. Death. He wanted to see her. He wanted to be dead.

Dierdre gently tightened her grip on Anna's hand in reassurance, in understanding, and for distraction. She was letting her know there was more to be seen and now was not the time to contemplate what had already been seen.

Suddenly they were back to the little yellow room, although this time, the man was approaching Dierdre. He was reaching for her and speaking although the words could not be defined. The tone was menacing and slow. There was a slight edge of familiarity to the distorted voice but Adrianna could not place it.

Dierdre's grip tightened further, though not in comfort. Her little brows were knit together in frustration as she concentrated. Vibrating, her tremors could be felt by all of them, drawing morbid curiosity mixed with worry from the group. As she concentrated deeper, the man's voice became clearer and it was now possible to see that his hair was short and jet black. Although his face was still blurry, it was handsome, boasting a stern jaw with full lips, and his eyes were dark.

The next voice heard was a total shock to the three observers, though Valerie was the first to gasp.

"What did you do, Daddy? Why is Nicholas not crying anymore? I want Mommy. Where is Mommy?"

"Your mother isn't home, child, and you will not speak to her. Nathaniel won't cry anymore, little one—he is happy. Now go on to your room. You will not say anything to anyone, Dierdre, or you will pay dearly, little girl."

His words were as clear as Dierdre's in their sound and meaning. He had killed the little boy, felt no remorse, and threatened to do the same to her if she were to tell anyone. They all looked at Dierdre, feeling her sorrow, loss, fear, and hurt. She was just a little girl, sweet, innocent, and mute. She should not have to bear witness to such brutality or evil. She lived what most only watched in movies, read in books, or dredged up from some dark recess of their mind when sleep consumed them.

The scene ended before their eyes, though the painful residue left behind would be permanently etched forever in their hearts. Dierdre no longer cried though she sat frozen on the full bench, having shared her deepest darkest secret with her companions.

14

Power and Realizations

There was nothing to be said. No words could minimize the atrocity they had all witnessed and Dierdre had lived. I tightened my grip on the pale little girl sitting meekly beside me. I felt fully conscious and coherent now. Jonathan opened his eyes once again, not seeing the ground, with its lush grass, or the waterfall before him. Valerie was the first to speak, her voice weak and lifeless.

"We cannot tarry here long. They will be coming for us. We must figure out what we are going to do. Adrianna, how do you feel?" There was no answer for this question, no description of the emotions that washed over us all; Valerie knew this and was really just setting the stage for me to be strong, even if I did not feel that way.

"I am okay, I think. I am not sure what happened, but I think Lucifer put me into some sort of trance or sleep to keep me here, although it was quite nice."

Shock showed on their faces as they realized that the dark angel had already shown his face here and that I had been alone and unprotected. Jonathan knew that it wouldn't have mattered had they been present, as they were all powerless against him, and that I and Dierdre were the only ones who had the power to stop him. When I thought of Dierdre, a lost memory came to the surface of my mind but sank back down just as quickly. There was something I needed to remember. Something about Dierdre and . . . the dark angel, but I couldn't grasp the memory enough.

"Oh my, what did he say? Did he hurt you? What does he want?" The questions came fast from Valerie as, surprisingly, Jonathan remained silent. I knew I must tell them what I knew and what I had decided, but I was afraid to hurt Jonathan and have them believe me to be insane for seeing some good in the darkest angel.

"I . . . He . . . Well, he wants to start the middle world over again. He did not hurt me though I was terrified. He wants . . . well, he wants me!" I felt Jonathan stiffen at my report; he all but vibrated with building rage.

"What do you mean, he wants you? Adrianna, we must know exactly what he plans to do so we can stop him." Valerie's impatient words cut through me, leaving me stung. Jonathan remained silent beside me, angry and rigid.

"Ok, please just listen to me; hear me out and don't interrupt. He was here with Juniper and Gloria. They are his helpers and have both been promised to be reunited with loved ones or something like that so they help him. They were the ones who banished the others we lost. He wants me to stay here and . . . crap, this is insane . . . well, to BE with him and rule the middle world with him. He thinks I am equal to him in powers and that I can help save others who wish to come here. He says he is trying to make right the awful things he has done and wants to be good like he was before."

I stopped, mostly because I realized that I was utterly confused about everything he had said to me, everything he hadn't said, and the thought that he had already been the most evil creature ever. But the idea of helping him become what he was always meant to be once again was too hard to let go of, despite what he had done. But I had no clue as to how I would convince the others. My memory of all that had been said seemed a bit foggy.

"That's incredible! You believe him!" The harsh words were a clear statement dripping with revulsion and rage. Jonathan glared at me as if I had just told him I was a mass murderer. I hadn't expected him to understand but I didn't want him to hate me for my feelings either.

"Jonathan, he knows what he has become and wants to change that. I am obviously not considering his proposition but I kind of made him believe I was. Regardless, I cannot ignore the fact that he wants to be good again, to change and to keep the middle world for others to be able to . . . ." Jonathan cut me off with venom in his heavy voice.

"NO, Adrianna, you sound like an idiot. He does not want to be good; he does not care for you. He is incapable of caring, or wishing, or dreaming like you. He is using you. He will destroy all of us and . . . ." As Jonathan had interrupted me, Valerie now jumped in to stop Jonathan's tirade.

"Listen, both of you. No good will come from you fighting over who believes what, although I agree with Jonathan, Adrianna; your compassion is overriding your common sense here. Adrianna, we know that somehow the only way to stop Lucifer from destroying us is for all of us to move forward: you, Jonathan, me, and especially Dierdre, as she seems to be his connection here. That is the only way, so I say let's do it right now. I can guide us through it."

She spoke with authority and conviction. I wanted to follow her, to believe she was right, to act on her command and leave, but I couldn't. There were things left undone. I had to find out what the vision of Willie meant. I had to know for myself that she and Jonathan were right about Lucifer and that he was just as lost as the souls he allowed to disappear, and I had to know what happened to Dierdre that brought her here. The memory that had touched my mind only moments before came rushing back. This time I concentrated and grabbed hold of the strings. Dierdre . . . what happened to Dierdre . . . her father killed her brother . . . her father . . . .

Ignoring the shock of Jonathan and Valerie, I grabbed Dierdre from beside me and sat her on my lap. Staring at me wide-eyed, a look of understanding and acquiescence crossed her precious face before me. She knew what I needed to see.

As the vision from earlier began again just where it had left off, I could hear voices in the distance. Women's voices. I wasn't sure if it was a part of the vision or not, but I was too absorbed to react. This time it felt like we were alone—Dierdre playing her story for only me, while Jonathan and Valerie had chosen not to see.

I watched as Dierdre ran from her father, back down the hall toward her room. She was followed instantly. Screaming at him to leave her alone, she slammed her bedroom door with a thud, which made me jump in spite of myself. A quiet laugh could be heard from the other side of the door as she jumped into her pink-sheeted and blanketed bed. The laugh was cold, disturbing, and authentically evil.

A new voice startled us both as the Dierdre in the vision quickly threw back the covers.

"Dierdre . . . Dierdre . . . Mommy's home, honey, where are you?" I knew in my heart that this was Dierdre's mom and she had a precious, loving heart. I hadn't thought much about Dierdre having a mother.

I heard footsteps quickly leaving the other side of the door, then quieter steps could be heard entering the room. I couldn't understand why the room was so dark, and I was unable to see anything in the room: walls, decorations, furniture, or even the little pink bed I knew she had climbed into.

I could see the door was wide open now, and a very attractive woman, holding a purse and coat, stepped lightly into the room. Dierdre's mother's brown hair, wide, soft eyes, and petite frame were a complete up-sized replica of the little girl on the bed. Dierdre leapt into her mother's out-stretched arms and began to sob as the dark shadow that had been creeping closer stopped just feet behind the embraced pair. I could feel the love between mother and daughter just as intensely as the evil looming in quickly.

"Honey, what is the matter?" Her mother's melodic voice rang through the room above the child's gut-wrenching sobs. I could tell she knew something terrible had happened.

"Daddy . . . Daddy killed him. Daddy killed Nicholas."

The shot was startlingly loud despite the added silencer on the gun. My body jumped in reaction, probably scaring Jonathan as well, although Dierdre didn't jump. I could feel that her eyes were closed and that she was crying, though I did not see this through my own eyes. All I could see was the lovely woman, clutching her only child, loosen her arms, and fall to the unseen ground while her blood covered the wall in front of her. Dierdre was tangled half under her dead mother's body when she stared up at her father. The screams erupting from her were not quite loud enough to drown out his final words or the last shot from the gun in his hands.

"I told you not to say a word."

Something was pulling me away, like a marionette being ripped up from the ground with a dozen strings pulling at the body and limbs. I wasn't ready; the vision did not feel done although I had seen more than enough horror for a thousand eternities. The pulling became stronger and the vision cleared just as some of the room was becoming illuminated. Dierdre hadn't wanted me to see her room—how gentle, loving, and cozy it must have been—then the innocence massacred by her mother's and then her own blood covering the walls and floor, but she was trying to let me see . . . something. Light came from behind the father, though even squinting my eyes I could not see any more than what we had all seen of the father before.

Then it was gone and everything was changed. Opening my eyes quickly, I realized I was cold. I was cold because I was alone. No one sat on the bench on either side of me; no one held my shoulders or my hand. I couldn't fathom how everything was so different and I hadn't been aware of anything happening. Hadn't Dierdre just been here showing me this vision? I couldn't have possibly seen it on my own. What the hell was going on?

The figure appeared before me just as the last remnants of the vision left my mind. As it faded into normal mental darkness I was sure, positive, that the face I had been waiting to see had become clear.

"How was your rest, dear Adrianna?" His voice was as familiar as if I had been hearing it for a hundred years, and oddly comforting. I was sure it was part of his angel magic that he was able to affect me so deeply by the very tone of his musical voice.

"I am rested, but confused; was I not just sitting with my friends?" I asked the question, knowing that he must be responsible for the change in atmosphere, and I feared for them. Where had he taken them? Had they been hurt? I knew I couldn't ask those questions right away if I wanted to continue the charade of believing in him, which no part of me did anymore. Now I knew the truth and trusted it, no matter how it confined and crushed my heart of natural instincts.

"You were, my dear, you were. However, they were needed back at the common room so I had Juniper bring them back. I am glad to have the time alone with you so that I may further explain some terrible things to you."

His voice was thick with sweet melody and very convincingly honest. Doubt tried to invade my confidence and shake my control. Dierdre's face tortured my mind and my poise returned. I would not fall for his lies.

"Well, Lucifer, it seems that we are alone and have the opportunity to chat, so . . . let's chat." The strength in my voice pleased me but the unmistakable edge of deception was there as well, though it didn't appear that he noticed. Or, if he did, he ignorantly swept it under his rug of always believing what he wants to.

His next words rocked my ship: the waves plummeted over the side, threatening to capsize me into the murky depths of despair. (I'm never stable.)

"Adrianna, I know you saw my face and believe it was I who murdered my daughter as well as that darling little infant and wife; I must beg you to know that I could not do such things." The shock was evident in my face and posture, and only my disbelief kept me from succumbing to the cold waters of my mind. I had taken on too much water though. With all that had happened and all I had seen this last while, I had lost most of the control that I needed to judge lie from truth, reality from deception, and good from evil. I could not be the angel they all thought me to be—I was weak, flaky, and falling apart.

"You see, Adrianna, I have certain . . . abilities . . . and I am able to control certain aspects of life and death. If I so choose, I can compromise the life of a . . . let's say . . . a shady character. Someone who has reached out to die through desperation or through evil deeds." I heard his words seep into my ears and reach my consciousness; I was becoming absorbed by his speech, though I tried very hard to fight it. I feared the manipulation he could unleash into my heart, oceans flooding my sanity.

"When I witnessed what that horrible man had done to his only son, I watched to see what else would unfold. I saw his mind decide to murder his wife and remaining child, if she were to speak of the tragedy. I could not intervene no matter how I wanted to. But I could reach into that room with my mind and grip the soul of little Dierdre as she was so mercilessly killed by her father. In turn, she became my daughter in this after-realm, and I her only father. If she were not mute she would tell you so herself. We are bound together by my saving her soul—that is why she is here, sweet Adrianna. You have longed for that answer, I am sure."

The doubt that had begun to stab at my resolve now tore it to shreds. He must be speaking the truth. It made sense, didn't it? Dierdre had been showing me visions to prove that he was not as evil as he once had been, that he had brought her here to save her soul. It was laid out, plain to see: the logic, the sequence of events, everything. But something didn't feel right. It felt like there was a festering sore inside my mind, pulsing with a heartbeat I no longer had, bleeding and oozing pus, only to be further rubbed raw with more information. The bandage would be the truth, the whole truth, and I could not give up until I found it, whether it meant my eternal end or not.

"Thank you for telling me this, Lucifer, although I knew there had to be an explanation. I know you could not have done those terrible things to those innocent people. You wouldn't." It was simple to see that I had succeeded in my dangerous ruse. The dark angel's face brightened and his eyes gleamed; he knew he had won. Or he believed he had. I wasn't so sure. Even his body seemed to relax and his skin glowed ever so slightly a faint purplish aura. Wings twitching, he came to sit effortlessly beside me, taking my hand in both of his. His massive wings positioned themselves to allow comfort, and his presence seemed to only heighten in intensity when he reduced his hierarchy by sitting at my level so close beside me.

"Adrianna, tell me, my love . . .," I cringed at his lips forming the phrase that only Jonathan had said to me, when it had caused my body to quiver in pleasure, "Tell me, what is it like to be the only female archangel to ever exist?"

It never ceased to amaze me how I could be so easily stunned into silence since I died (I used to always be quicker with comebacks); his mystical voice and incredible comment had me mute. Again, I felt weak, though not rocking at sea or plunging into the icy waves—this time I felt like I was standing on top of a large cliff, bearing my first set of wings but too afraid to jump, believing they would not give me flight.

"How . . . how can I be? An archangel? I am just a stupid girl from Canada." My retort went unnoticed.

"I wonder what the mortal world would be like if there were more selfless beings like you. I am not sure how only you have come to be what you are, Adrianna, maybe for me to finally see the light of good again and atone for my mistakes—I like to believe that, of course. No matter the reason for your existence, here you are. Full of powers you can only imagine: the power to see, hear, touch, taste, smell, and manipulate what others cannot conceive of. It is truly amazing. In time, I will teach you, show you what you can become, with my help, of course. But now I have pressing matters that need my attention and I would request that you join me."

I began to panic as soon as the word "join" escaped his lips. I would not allow myself to become arrogant or big-headed with his tale of my royalty. If I was indeed this female archangel (I was definitely a girl), then I would have to quickly learn how to use enough power to prevent him from harming or getting rid of anyone. No matter how his explanations, stories, and truths nipped at my heart, forcing me to believe everything he said, I knew something was still wrong. He was still pushing an even bigger façade onto me, as I was to him. I knew I must stay strong with that knowledge, because if I didn't I would be responsible for what happened to everyone here I had grown to care about.

I couldn't decline to accompany him, as it would make him suspicious and possibly angry, so I slowly stood up from the bench and walked a few steps forward, my back to him. I did not hear but felt him stand as well and step towards me. His large but beautiful hands rested on my shoulders, caressing the base of my neck, then my arms. His touch gave me goose-bumps out of a mixture of pleasure and fear. Needing to stop his touch from affecting me the way it did, I turned quickly to face him. Surprise and disapproval were evident in his eyes at once, so I quickly gave him cause to change. Closing my eyes, flinging my arms up high in the air, I smiled and laughed with as much false but convincing enthusiasm as I could muster.

"So, when do I get my glorious wings? Shouldn't all special angels have them?" A startling music filled the air around us; the beautiful tune caressed my ears and I opened my eyes to see it was Lucifer laughing. He looked completely happy and joyful, light, and pure. It tore my heart to know I would have to try to destroy him if it was in my power to save my friends. Struggling to keep the moment going, I spun around once and flapped my arms.

"Maybe I will become the most powerful and beautiful angel ever created." I looked up to see the reaction I had caused, curious to see if it would make him angry to think I conceived of the notion of being more powerful than him, but I did not understand the reaction I saw.

Everything happened so suddenly, like a bomb exploded and the world changed again, like earlier. This time, voices invaded my mind and I felt surrounded. Then the pain came.

I at once fell to my knees, desperately stuffing down the scream that was building in conjunction with the throbbing surfacing in my head. Pressure, mounting and threatening to burst, seared through me. I tasted blood in my mouth and realized I had chomped hard on my lip. Was he was doing this to me? Had I upset him that much with the humorous display? Was he taking back his control, showing me without a doubt that he was stronger and more powerful than I would ever be? I felt defeated and more terrified than I had since I arrived at this very spot or this very world.

As the terror and throbbing grew simultaneously, I heard footfalls and voices approaching; the chaos that had transpired before I fell was becoming clearer and closer. Who it was that dared to interrupt this game of supremacy I did not know, but I was quite sure it would be no one to rival this force that had so easily diminished me. Before the blackness took over and enveloped my mind completely, I thought for sure I caught a scent I knew well, a scent I had grown to love. Agony ripped through me in one final treacherous ripple as I realized Jonathan had come to save me against an enemy that could and would demolish him with a snap of his angelic fingers.

* * * * *

"Do not touch her. What have you done to her?" Jonathan yelled before bridging the distance between himself and the dark angel. Following close behind was Valerie, now clutching a tiny Dierdre in her arms. Jonathan saw Adrianna lying on the ground at Lucifer's feet; she appeared to be unconscious. Reaching her side, he crouched down, ignoring the towering lord above him. Gently, he took her into his arms, her head flopping into the crook of his elbow. She was pale and slightly cold but still her chest rose and fell needlessly. It still amazed him how the human form retained its instincts even long after its death.

"Anna, it's me." Her eyes fluttered at the sound of his voice, pulling her slowly to the state of consciousness. Her mouth parted as if she was about to speak, but no words emerged. Anxious waves rolled off her, terrifying Jonathan even further. He wanted to keep holding her but the inevitable confrontation awaited. The angel stood just above him and Anna, smiling with his full lips in a grin that didn't reach his eyes. In the depths of the blackness where his regular and spiritual vision came, there was anger and contempt.

Jonathan felt Anna move in his grasp, though he did not realize she had sat up until he tore his eyes from the dark angel's hardened face. He wasn't sure but he was almost positive he could see a small spark of fear and maybe jealousy mixed with the hatred. He stood, taking a few steps back from Anna and the dark angel to gaze into his malevolent eyes. At once, Jonathan saw in the dark depths that Lucifer believed he owned Adrianna, and that Jonathan was interfering with his plan to dominate her.

"Well, if it isn't the only suicide not to end up lost. It amazes me that you have befriended my Adrianna, and she you. I have wondered how you escaped that room. Is it because you are special, Jonathan?" It was not a question he wished an answer to. Lucifer's satiric smile was cold and full of his conceit. It was plain to see he wanted nothing more than to crush him like a bug, and the fact that he WAS the only suicide not to end up mumbling to himself in a room full of like individuals for eternity was a mystery to all of them. He was very thankful that Adrianna had never had the courage to ask what had been his demise. He was embarrassed to tell her. But now she knew.

Jonathan felt no more significant than an ant in the dark angel's gaze, and he knew he was just another an enemy the dark angel would conquer. They all were. She was the reason he was here—she was what he sought. He also knew Adrianna knew this already, as she had begun telling them earlier. She was aware of his intentions and would have a plan based on this. This scared him even more. If he coveted her, he would stop at nothing to possess her, though if she could keep up the façade she was no doubt presenting, she would be safe until he could find a way to protect her.

"Gloria." He did not yell, but the intensity in the dark angel's voice reverberated with a power which none could defy. "Come here, my child."

The word "child" sounded quite amusing being used to describe an overweight woman who was obviously as old as the hills, but she obeyed with trepidation. With her ever-obedient Juniper at her side, she walked slowly from behind Valerie and Dierdre, pausing only momentarily to glare evilly at the two scared figures.

"I have given you certain gifts, have I not, Gloria?"

His question confused everyone, including its addressee. Gloria nodded in answer but appeared too afraid to speak.

"I gave you these gifts to aid me in my quests; however, I did not grant you permission to use them in my presence to interfere. Did I ask you to hurt Adrianna?"

"No, you did not, but she was going to attack you so I thought it best to stop her." She replied confidently, as if this excused her interference. Adrianna opened her eyes as the obvious relief from the pain relaxed her body and brought her to complete consciousness. Jonathan was pained to see her look up at the dark angel in astonishment and contentment that it was not he who had hurt her. Her gaze then shifted to Jonathan with worry and regret plain on her ashen, tear-streaked face. How he wanted to hold her again. She looked so fragile, still sitting on the ground.

As soon as he had thought it, the angel took his desired action away from him. Crouching down, holding out both of his strong arms, his wings opening slightly to give glamour to his graceful movement, he scooped up Adrianna, holding her tight as he gently placed her on her shaky feet at his side. One arm remaining around her waist, he leaned in to her ear, telling her something that was only for her to hear. As soon as he turned his head back towards Gloria, a small sound of horror escaped from Anna's mouth as she covered her ears and shut her eyes. But it was too late to have any effect.

The dark angel's wings rose part way and opened in a whoosh. A purple glow emanated from his body at the same time as he snapped his fingers in front of Gloria. The large woman faded to the mere form of a ghost in seconds, a look of complete terror and disbelief on her round face. Then she was gone completely.

Juniper ran forward, yelling, but stopped short when the angel raised his hand like a crossing guard telling the vehicles it was unsafe to drive. It was very apparent she would suffer the same fate as her comrade if she were to advance or even speak to him about what had just transpired before them all.

Adrianna again looked at the dark angel with more ease than Jonathan wanted to see. In her face was sorrow for what had happened, fear enough to keep her from saying anything about it, and apprehension at what was left to come. He longed to kiss her soft mouth, his own twisting into a grimace at the thought that the dark angel wanted to touch her too, was touching her now and giving her comfort in her sadness over the murderous deed the dark angel himself had just committed without even a hint of remorse.

The dark angel turned slowly towards Adrianna's ear to once again whisper sounds for only her to hear. Her expression softened noticeably, though there was residual hardness that did not leave her vacant eyes. Jonathan wanted desperately to believe she was not being manipulated again by this coercive figure of evil, but he knew her, better than anyone knew her, and she was optimistic to a fault, believing there was good in everyone, no matter how impossible the notion. Jonathan knew that there was no good left in this dangerous creature; there may have been once, though it remained no more.

With astonishment, Jonathan noticed that as Lucifer spoke his treacherous words to Anna, his skin began to shimmer. It wavered as if it were just a hologram, fading as Shadow had before he left so suddenly. This could be good news. Very good news. Putting the facts together of the times Lucifer had visited here, his connection with Dierdre, and the use of his powers only moments ago, Jonathan easily deduced that his power seemed to diminish both the longer he remained here and with each direct use of it. He fervently wished he could relay this to Adrianna. Glancing at Valerie, he realized that she had come to the same conclusion, but Anna seemed to remain oblivious.

"It seems we have a rather unfriendly gathering here." His statement came out of nowhere and everyone jumped. Dierdre cringed into Valerie's neck at the tone of her father's voice. Anna did not miss this reaction, nor did anyone else.

"I was just having a few private moments with Adrianna and you three were supposed to be dealing with some . . . upset in the common room, I believe. I will have Juniper take you back."

There was urgency in his voice that had not been there before. Desperation shot from his eyes as he realized that no one was heeding, or would heed, his orders. Clenching both fists, he slowly let go of Anna and walked a step in front of her. Jonathan reciprocated by stepping forward as well. The plan that had been formulating in his mind became clear with his advance toward the enemy. Praying Valerie would also understand his reason for his insane reactions, he advanced again.

Instantly, the dark angel lifted his hand and Jonathan dropped to his knees. The pain was intense but not enveloping. He had felt and withstood worse, although not since he had been a mortal. Suppressing his agony, he concentrated on Anna's reaction, which was just as he prayed it would be.

"Stop! What are you doing to him?" she yelled viciously. Walking closer to Jonathan, Anna turned to face the lord before her, his powers instantly dissolving, the hold broken. Jonathan stood but stepped back to regain his faltering composure. As expected, Lucifer began to fade just a little more. This time Anna saw it. The words out of her mouth surprised them all.

"Please do not hurt him, Lucifer—look what it is doing to you," she cried. The dark lord stopped short, looking down at his disintegrating form, and Jonathan nearly gagged aloud at the worry in her voice. Incredulously he stared at her beautiful face, her intelligent face, and was startled to see her wink at him. She knew! She was letting him know that he was right in his quest to diminish the angel's powers so that they could make their escape. He had never loved her as much as he loved her at that very moment. All that transpired between them was lost on Lucifer as he continued to stare at his own fading figure and then at Adrianna. Fear was plain in his face though the strongest expression he wore was one of triumph. He still believed completely that Adrianna was his. She would give herself to him and they would rule the middle world together for eternity. Her ploy had worked, and that, coupled with Lucifer's inability to accept defeat, meant Anna had won, for now.

"You need not worry for me, sweet Adrianna. I merely need to attend to a few things for a short time. May I ask that you rest once again while I take your friends from your presence?" It was her final test, of that she was sure. She had no choice but to agree, but how could she know they would be safe? How could they find their way to conquer him and move forward? Knowing they must saddened her deeply and she knew she must save them.

"I will rest, but . . .," standing up on her tiptoes, she leaned into him and whispered the last of her sentence in his covered ear. A slight frown of worry knit his eyebrows but cleared quickly when she kissed his cheek.

"That is the way it will be. Rest easy, Adrianna, and I will be back for you."

As the words left his lips, the ground seemed to move beneath them. A flash of strong purple light, full black wings, and a wisp of a black robe was all there was to see. Jonathan stared intently at his new surroundings where he, Valerie, and Dierdre stood. Adrianna was nowhere to be seen. He had separated them before he left, possibly at Adrianna's command. Jonathan feared that she may have been placed in a trance once again and that they wouldn't find her before Lucifer regained his power. He wondered how much longer Adrianna could keep up with her charade before the dark angel asked her to do something she wouldn't do.

Jonathan also knew one other fact that he had been careful not to concentrate on. In a very short period of time, he was going to die. He knew they must move forward and he would lose Adrianna. He also knew that being without her was worse than any death, his own personal hell. Though he could not be the man she deserved, he would gladly stay near her, even if she found someone more deserving than him, to see her every day, to give her all he had and find ways to give more. The need to know her—just to know her.

As each moment with her passed, he gained more insight into how wonderful she really was. Also with each moment, he realized he had never given her enough. He hadn't shown her he loved her every minute of every day, he hadn't put her feelings above his own in each situation they had faced, and he hadn't fought hard enough to be strong for her when she needed him to be. He had failed her and now he was about to lose her.

It was a huge pill to swallow, the realization that he had taken for granted even one second of the time she had given him. Even the times she scolded him, was angry at his stubborn ways, was irritated at one thing or another. Those times especially he should have honoured, because she cared enough to be hurt when he was distant, cold, or just plain foolish. She loved him enough to want to help him become a better man for himself and for her. She gave him every ounce of goodness he knew and he never appreciated the gifts she really gave. Selfishness had ruled his life and his death. Immaturity, narrow-mindedness, and the fear of growth, change, and responsibility had killed his mother, taken his sister and brother from him, taken his own life, and soon would take from him the most wonderful person he could ever hope to meet and love.

It was his past, his present, but he would make sure it was not his future (if he had one). If he had one hour or a million left, he would use every second to honour her and her love. He would give all he had. She would know he loved her and would always love her.

15

Sacrifice

I knew I was resting as the heightened physical sense of the atmosphere was very strong, though I could not *see* my surroundings. I knew my plan had worked. He was gone. Gone to replenish his strength, to honour the bonds placed on him so many eons ago, to fall back into the mortal world as a fallible, mundane being. (I bet he must hate that). I knew I didn't have much time but I wasn't even sure I could do what I was planning to do. I had managed to stay coherent this time as he put me into the sleep-trance but I had to find a way to be fully conscious and get to the common room as quick as I could. If he came back, fully powered up, he would know I had deceived him and we would all surely die our final death.

Drawing on an internal strength I was slowly realizing that I had, I forced my eyes to open, to stare intently at the grass, seeing each blade curve and intertwine with the next, the vibrant green against the pale blue above. The breeze felt warm and soft on my skin, carrying a scent similar to orange pekoe tea bags; I wondered where that would be coming from. Taking in every detail of my surroundings cleared the remaining fog from my head and I knew I could stand.

My legs felt wobbly at first, as if I'd been doing too many squat reps (great but painful way to tone the butt). I quickly got my ground and found myself almost sprinting towards the path to the common room. The urgency began to build with each step and I longed to see their faces, to know that the dark angel had not harmed a single hair on their precious heads. Although I wanted to see Jonathan the most, to apologise for the hurt he must have felt at my deceptive words and actions to Lucifer, I felt an acute need to hug Dierdre, to give all the comfort I could for all she had lost and all that she had shown us. This gave me an extra boost and I all-out ran as fast as my newborn legs could carry me (I probably looked like a week-old fawn scampering along).

With every step, I prayed there would be enough time. I had to save them or an eternal death was all I would wish for and deserve. As I focused on my path, a terrible thought stabbed through my head. I realized with astonishment that no matter how this turned out I would be saying goodbye to these three people I had grown to love, including the one who had my heart. My soul mate. I would be bidding farewell, hopefully with all of us moving forward to live again as mortals. Maybe, just maybe, if we were all bound as Valerie thought we were, we would meet again in another life. I prayed this was true. But if we lost and our souls were destroyed, then all was lost forever. I would never feel the emotions that touched, invaded and encompassed my heart from those who had made me believe in myself and love.

Tears streaming down my face, I ran blindly towards the shadow of the room up ahead. If I hadn't smelled the scent that had frequently absorbed my senses I might have run right into him. I stopped dead in my tracks, almost tumbling over in the process. His arms reached for me protectively and steadied me. For what seemed like a very long piece of sweet time, he stared into my eyes until I couldn't stand it anymore. We had such little time.

Lunging for him, I wrapped my small arms around his strong shoulders and neck, my fingers finding his hair as his lips found mine. I didn't breathe. He was my breath, he was my substance of life, and he and all he knew and believed were what guided me, what gave me my gifts. Our lips felt fused together, our bodies one, whole, complete, sure, everlasting. I didn't want to let go. Ever.

When Valerie cleared her throat, I knew it was time. Time to break the embrace I needed so much. Time to do what I knew deep inside I could do. I could save them. I had to. I wanted answers to my questions (maybe just one or two of the million).

"Jonathan, how are . . . I mean . . . did you kill yourself?" I instantly regretted asking in front of an audience, though the lack of time we had freed my tongue.

"Time is against us, my sweet, inquisitive Adrianna. Remind me, and I promise to answer your question with a tale someday." I opened my mouth to protest the absurdity of his words when he raised hand and a look of impatient fear replaced the softness. Releasing Jonathan left me cold and torn inside but knowing he felt the same gave me comfort. Soul mates. I glanced at Valerie holding Dierdre again, and was delighted to see a careful but honest smile on Dierdre's angelic face. I returned the smile and reached for her. She was like a fragile infant with intelligence, strength, and beauty that surpassed us all. I was stunned when she would not come into my arms, her face breaking the smile into a frown. She had appeared so happy to see me—why would she not want me to hold her? Comfort her? Maybe she was still too raw from all I had seen. I couldn't think about that now.

"We did it," I said, not knowing how to tell them we must say goodbye, but knowing somehow that they already knew.

"No, Adrianna, you did it. You are amazing. I won't ask how you managed to get him to send us here, to use the last of his power to do so, and how you could avoid his trance. Incredible."

His words were tender and flowed into my ears and heart. I could listen to his voice forever. As it had been my salvation when I first arrived, and from that day forward, no matter where I was or even who I was, his voice would always be remembered.

"There is no time to pat each other's backs now. He will not be long in regaining his power, and we must move forward. Adrianna, I will show you how it is done and you will hold each of our hands. We will go together." Valerie, ever the voice of reason and authority, spoke the command we needed to hear. But there were things left undealt with, left unsaid.

"But I cannot just leave—what about the middle-worlder's? Juniper is still here, she will help Lucifer when he returns, and they together will destroy all those souls. I promised them. I cannot just leave them." The panic took over before I could rein it in, and the tears flowed again. Jonathan reached for my hand, holding it tight. There was no electric shock as there had been many times in the past, but there was comfort. Comfort I desired from only him.

"Adrianna, we have figured out that Dierdre here is the only reason Lucifer is able to be here in the middle world again. We don't know why this is or how it works, but when Dierdre moves forward he will not be able to come here. He will be bound to the mortal world until he finds a way to get in here again. We need to move forward now; Dierdre must go with us for it to be over." Jonathan spoke clearly to ease me slowly, to allow no doubt in his words. I knew he was right—they would be safe—and I also knew I could not say goodbye, there just wasn't any time. Seeing all their faces, Hank, Maria, Peter, Donald, so many precious souls, scared souls, who would be safe soon.

"How do we do this? How do we say goodbye?" This time the tears were accompanied by wrenching sobs. I knew this was coming, but I was not prepared. Holding on to Jonathan's hand was not enough—I wanted to be swaddled in his body, and I wanted to hold Dierdre, and even Valerie. I couldn't find the strength.

Jonathan took his hands and placed them tenderly on each side of my wet face. Leaning his forehead onto mine, he spoke, his words caressing every part of my body and soul.

"I have never lived, Adrianna—my years in the mortal world, my time here . . . I have never tasted life—until you. I was no more than one of the lost souls, the forsaken; I caused pain, felt pain, but never knew love, kindness, or faith until I saw inside your soul and your unbelievable touch on my heart. I want to move forward, taking the torture of our time apart, knowing and believing beyond the shadow of a doubt that we will meet again. Adrianna, we are meant to be, we are soul mates, and we will find each other again."

I don't know when I stopped crying, I don't know when I ended up on his lap, in his safe arms, but there I was, enthralled by his words of truth, mesmerized by the sincerity in his gentle, loving voice. I could move forward—there was no goodbye. Not trusting myself to speak, I stood slowly, placing a kiss on his forehead as I stood. I reached out for Valerie, at first still unsure whether Dierdre would accept my touch, and embraced the woman that I had once doubted but always admired.

"Valerie, I will always know you. I don't know how, but I will. I hope we meet again. Thank you for everything." She smiled and I was surprised to see tears gleaming in her aged, intelligent eyes. There was something else there as well. A secret she had, or knowledge she hadn't shared. I briefly worried she might have "seen" something she hadn't revealed. I didn't ask.

Turning toward Dierdre, I reached for her hand. Once again, she subtly shied away from me. Hurt flashed through me as I pondered the little girl's reaction once again. What had I done to give her cause to reject me? She bore no look of anger or hatred, no fear emanated from her, just sadness. Maybe she feared breaking down if she accepted my hand. I hoped beyond hope that I hadn't hurt her.

"What must I do, Valerie? It must be now, before I cannot leave." I grabbed her hand with the one of mine not being held by Jonathan's warm, gentle hand.

"I obviously haven't done this before, but Shadow did walk me through the steps verbally before he left. It has to come from the desire to leave, and the belief that you deserve to move forward and be born again—not from greed or ego, just the desire to live again to do good. To be good. Adrianna, there is no one more deserving than you." She smiled and squeezed my hand.

"You must be touching each of us, preferably holding both Jonathan and Dierdre's hands. Your power will send them on, no matter their thoughts. I don't know how it feels or what happens after that. I wish I knew but I believe it will be free and painless. I pray for all of us." There was no more to be said. If I didn't go now I never would. Jonathan smiled at me, his eyes teary, frightened, but sure. I felt powerful in his gaze, felt like everything would be all right.

"Ok, let's do this!" I shouted the words, the force of them setting us all on course—to what end, we did not know, but we had faith.

Out of the sky, the clear, liquid sky above, came the sound: the torturous, malevolent sound, piercing with suffering and wrath. Dierdre's little body shook violently and her mouth opened in a silent scream. The sound ripped through the air like the breaking of the sound barrier, and I knew that Lucifer knew. He could not materialize, but he had heard, and if he had even an ounce of power left in him, he would stop us and obliterate us all . . . .

Now was the time. I tightened my grip on Jonathan's hand and grabbed for Dierdre's. This time she had no choice but to accept, as I was not taking no for an answer. Valerie stood inches behind me, appearing to be ready to grab hold when the need arose. I quietly wondered why she hadn't taken hold already, the reserved, sad look on her face intensifying my worry.

My hand was just centimetres from Dierdre's when the flash occurred. At first, I thought that Lucifer had pooled enough strength to appear, and I looked to the sky for wings; Jonathan followed my gaze as well, thinking the same thing. But there were no wings and I did not clearly see the sky—there was a room in the way, a dark-lit, rank room. Our fingers touched and the vision became so strong it rocked me back, almost breaking my connection with Jonathan and Dierdre. I held firm and closed my eyes. This had to be seen, but there was no time. Lucifer was probably on his way to crush us but it didn't fade, it didn't waver; it increased in severity as each second passed.

Valerie's voice broke through the silence of the scene before me like a song through water, far away and gargled, but the message was clear. We were out of time.

Not knowing how to break the vision without risking losing Dierdre by letting go, I held strong, knowing that there must be a reason this was happening now. There were lights coming from the room now, and a few dark figures stood near a metal-framed table. The wallpaper hung in filthy strips trailing down to the cracked, worn, grimy floor. An orange chair sat in the corner, unusable, having not seen its stuffing for many years. No other furniture or decorations adorned the room but for the four figures at a metal-legged table.

One figure sat, head down, concentrating on tying a rubber band around his arm. I instantly knew from the last similar vision I had had that there would be a syringe close by and that Willie would be the one in the rickety chair. Two of the figures left, shaking their heads, laughing in a slow, sinister way. I couldn't make out if they were laughing with any feeling or if it was the laugh of disbelief one gives when having seen or heard something incredibly senseless and terribly wrong.

One man remained, back turned, in front of the man in the chair. His long black hair came cascading in soft layers from a blue ball cap; the leather jacket and designer jeans he wore gave the impression that he was too good for the scene before him or that he would profit greatly from it. I tried to turn physically, as if it would shift the scene for me to see the figure in the chair, but both Dierdre and Jonathan held tight so I couldn't move much. The positioning finally changed and what was before me, before all of us, was sadly what I expected to see, matching with the words I could now hear coming from the sitting man's mouth. I knew what was happening.

"Will this be enough to kill me instantly?" Willie's weak voice filled the room despite its being barely a whisper. I was concentrating on his words so completely it could have been shouted through a megaphone.

"This will be enough to do the job, boy, but be sure it's what you want, because no amount of medical intervention will bring you back once this goes in. Do it up, boy—you seem to have your mind set, and who am I to stand in a determined man's way?"

The voice was off, but the same: not younger, not older, just altered. There was no musical tone, no gentle waves, no olden-day dialect, but it was him: it was Lucifer's voice. So, this is how he will stop me. This is how he chooses to win the game. I wasn't surprised, exactly; I am not sure how to describe what I felt when I knew he was there to make sure Willie would die by suicide and inevitably be lost forever. Lucifer knew I would be a tortured soul for eternity if that were to happen to Willie, that I would join him just to save willie from the lost room. He also knew I would try to stop him first, which would give him time to gain the strength he needed.

Suddenly it all became clear to me. Dierdre knew this was happening, or would happen as soon as the dark angel felt my final resolve. She avoided my hand so I would not see; she knew I couldn't leave with Willie dying. She knew what I would do before I knew myself.

But it was decided. I had to go. I had to push with all I had to reach into the living world and stop my friend. My best friend. I could not know his fate and walk smoothly into the void of unknowing. I instantly jerked both of my hands free of their innocent captors and shoved with my mind. I did not speak this time, I did not fear, I did not hope or pray—I became. My body heaved and was gone, flying weightlessly no wind, no sound, no color. I touched the ground quietly, not making the occupants aware of my presence. Slowly, I walked up to the table behind Lucifer and for the first time in my life I wished I held a weapon, shocked by the knowledge that I would use it for the kill.

There was no pleasure in my success in materializing in this room, no sense of victory that I was here just feet away from the only other boy I had ever loved besides Jonathan, no satisfaction in the sudden abilities I was finally utilizing. Despair, confusion, and desperation gripped me with speechlessness and almost brought me to my knees.

I could not give in so I pushed forward this time bridging the distance between myself and the two men: one whom I so desperately wanted to save, and one I would do almost anything to annihilate.

The reception I got when my appearance was known would have been almost comical under any other circumstance. I expected shock from Willie at seeing his long-dead best friend before him, seconds before he was to end his own life, and I expected anger from Lucifer for my deception and my final attempt to foil his plans. The dark angel just stood smiling at me, his eyes dark and menacing, and his lips thin, white, and firm. There was no humour in his eyes but he still believed he had won, that Willie would prevent me from returning to the middle world before he himself could return and destroy my other loved ones. He was so sure he had won. And win he just might, but I would not go down without a brawl.

Where the urge to grab his hand came from I did not know, and was kind of sickened by, but I found my hand wrapped vice-like around his fingers. Anger clouded his face when he realized he was unable to pull away from me. He was stuck, and I certainly didn't know how in the hell I was keeping hold, but it was happening. Slightly smug, but afraid, I turned my attention towards a very bewildered but delighted Willie.

He was not the boy I had left; he was hard, cold, and pale, bearing new scars, bruises, and circles under his eyes not derived from the same cause as before. He had become a faithless, drug-addicted murderer, but I loved him—loved his soft eyes, his tall body, the beautiful lean face that I had memorized over so many years, and the true heart that I knew was once a connected soul. He stunned me by speaking first, like he was not surprised I was there at all.

"Adrianna, you look heavenly. I haven't even pushed the shit in my vein and you are here. I knew it would happen. I knew I would see you again. You came for me.". His happy, dreamy words of delusion were like daggers being thrown into my body, each one a direct hit to a vital organ. I could almost feel the blood draining agonizingly from my carcass, my life pouring onto the floor. Lucifer's hand was slipping from my grasp, my powers weakening with my mental blood loss. I was not resilient enough; I should have known. This was it. I couldn't survive losing him, nor could I survive losing Jonathan. I had no choice but to choose who to save. If I stayed here and let Lucifer go, I might be able to save Willie from injecting the lethal venom into his arm, but then certainly Lucifer would have the power to return and eternally destroy those I had selfishly left behind, and they all would be lost forever. If I left Willie to die, I was confident that the torture would haunt me through any lifetimes I could have, even if I could attempt to move forward, which I knew in my heart I could not. There were no options. I had doomed us all. Lucifer had indeed won.

Tears flowed down my face as I stared at my delusional best friend. The boy I knew was a whisper inside of a shell, but it was then I could see the low flame deep in the recesses of his eyes. I felt my body walking towards him, knowing there was little point but to touch him once, one last time, a chance I never dreamed I would have but wanted so dearly. Reaching out, I caressed his face, feeling his own tears flowing soundlessly. His body shuddered beneath my hand and vibrated more when I leaned in to kiss his cold forehead.

I could feel Lucifer slipping further away the closer I was to Willie. I had nano-seconds to decide. But, I had already chosen. I needed to accept . . . and fast. (Never was I good at letting go.) My lips grazed his ear as I whispered the words I knew I must say, my dagger wounds searing with heat, torturing my every cell.

"I love you, Willie, but I cannot save you. I am eternally sorry."

Confusion crossed his face again and I knew he did not understand my words. I was also quite sure he would be dead within a matter of minutes after I left. I could not change what was inevitable and I could not allow Jonathan, Dierdre and Valerie to be harmed because of my weakness, my inability to save all I loved. I had to go back.

As soon as I made the decision to leave, I felt my chest fill with pressure and a blanket of force come down from above me. I was going to implode and explode at the same time. As suddenly as it had begun, it ended, and there was another in the room. At first, I thought one of the other men had come back to clean up the mess; the corpse that once was my dearest friend, but then I allowed my eyes to register on the man before me and what he was doing. Only the heavens know whether I could have moved if I had wanted to, or what I would have done if I could, but I will never know. The swiftness of his movements was disorienting, as was the graveness of his beautiful face. Like watching a movie in an empty theatre, I saw Jonathan stand behind Willie, lean over with both of his hands, and depress the syringe full of death into his waiting arm. Willie's eyes closed almost instantly but the light smile I loved stayed on his cold lips. Time slowed as I watched him die in front of me by the hands of the man I loved with all my heart.

Jonathan finally looked up at me, pure dread on his face. He believed I would hate him for what he had done. I wondered how I was supposed to feel, to react, but strangely, I didn't feel anything but sadness. I decided there must be anger there, floating inside my head, but it wasn't ready to surface yet; it would, I expected, when I allowed it to invade my mind. I reached out my hand at the same time as I heard and felt the terrorizing anger erupt from the dark angel behind us. He was free of my grasp but couldn't touch me. I was only here in spirit—I couldn't be harmed. He would leave, and I couldn't stop him. I had to reach the middle world first.

The floating feeling was slower this time, but I was not alone. Jonathan held on to me as we left my dead best friend behind. I glanced back at his empty face, amazed to still see his lips displaying the most enchanting grin. I would miss him with all my soul. I had no time to cry. Not knowing if Lucifer had regained enough power to return gave me the incentive I needed to move as fast as I could. I would mourn for eternity, I knew this.

We returned, still hand in hand, directly to where we had left. I was amazed at how easy it was to transport myself back and forth and wished fleetingly that I had more time to do it again. Dierdre was the first to move, jumping into my arms without a hint of the earlier hesitation to touch me. Her hug was fierce and warm. The tears that had been ebbing flowed naturally now and I wondered how one person, dead or alive, could cry so much.

Jonathan had let go of my hand and stepped back, his face pallid and filled with a sorrow I wasn't sure I understood. Valerie too wore an expression of torture that told me she knew what had happened in the mortal world and understood the ramifications more than I. Shrugging off the desolate feeling they seemed to project, I hugged Dierdre tighter and spoke in her little ear.

"We must go now, little one. You understand you are free to move forward and to be born again."

Her brows knit together as she shook her head slowly. I was astonished at her reaction. She didn't want to go. I couldn't fathom why.

"We must go now," I said gently but with urgency. Surely, she understood that he could come back any second and all would be lost. "Dierdre, please!"

"She feels like she has to stay, Adrianna; she feels responsible." Valerie's words were void of emotion. "She must go, but I must not. I have lived enough, and I want to remain here. I will find Juniper. Things have changed—you must take Dierdre, now!"

My head whirled with emotion, fear, and too much information to process. Valerie wasn't going to move forward. Dierdre felt responsible. For what? I had a fleeting thought of my brother, still lost to everything; I was glad that Valerie would be here for him.

As the questions flowed through my mind I felt the familiar current, not sharp and uncomfortable as before, but as a soft vibration, like the first few times Dierdre had projected her feelings to me. Suddenly I knew, I understood. The child hadn't spoken because she believed her voice was what had caused her mother's and her own death. She blamed herself. Ache soared through my veins again as I wished, with all my heart, the next lives we all lived, would not be of torture, pain, and loss. I had already had enough for a thousand lives.

"Dierdre, your brother was murdered by a vicious, evil madman, the darkest of evil spirits ever created. He also murdered your mother and you. It is not your fault. He would have killed you both whether you spoke or not. It might not have been that day, but it would have happened. He needed you to come here. He wanted you to feel this way so that he could use you to gain entry here. When we move forward, Dierdre, he will not be able to come back."

Her little green eyes widened exponentially, her mouth parting silently. She had not known. Although she was still confused, ashamed, and scared, I could see she felt better—a burden she had carried for so long had been removed from her tiny frame, and although the residual damage would stay, she could move forward knowing she could learn to stop blaming herself for what had been beyond her control.

She nodded her head in response to my original statement, agreeing quietly that we must go as if nothing had transpired since. I was relieved to see that although she was still tortured by her past, she would be okay. We all would.

There was electricity in the air that was not present before. At first, I thought it was because I was getting ready to move forward and be reborn. Goosebumps rose on my arms and I felt a cold sear of pain stream up my neck. I knew then it wasn't me causing the change in our atmosphere: it was Lucifer . . . he was on his way. I must act now. Without another thought I began setting my mind toward the steps ahead, which was the easy part; not looking back was much harder.

Dierdre's hand was tight in mine as I focused on my thoughts, concentrating on freeing myself from my bonds to this world, detaching myself from my past life to make way for the next. My mind was clear. I would be me wherever I went, I would live again. Dierdre would be reborn to a life without the chains that bound her here, without her pain.

I knew that the process had started and could not be stopped as soon as I felt myself float. No, I wasn't floating, I was flying. The emerging wind fluttered through two of the most beautiful wings I had ever seen, miraculously attached to my back, as my white robe whipped around my naked feet. The faint feel of silk caressed me as my raven hair flickered around my shoulders. My precious gift, the amethyst amulet, safely around my neck, was glowing. Dierdre was gazing at me with immediate acceptance - she had known - she had drawn this very image.

I shook with dread when the drawing surfaced in my head, I whipped around to stare behind me, knowing instantly what I would see. Jonathan knelt on one knee, holding his head in his hands. Torture was rippling through him, his burning tears cascading down his agonized face. He hadn't taken my hand. I had left him.

I felt my wings shiver violently, my power diminishing, I couldn't hold on, though I had to. I felt a tug at my hand and realized Dierdre was staring at me. Her mouth opened and the most amazing sound emerged, weak but harmonious.

"He can't come. By ending Willie's life, he gave up his own. Willie will now move forward, as he is not a suicide. He traded his chance to move forward to save Willie and you. His last act was to be selfless and worthy of you. Jonathan cannot follow."

Her first words spoken were ones of immense maturity, of wisdom, understanding and astounding power, bringing a message that chilled my dead, un-beating heart. The last I would hear. Her smile was radiant and comforting as we thanked each other with our touch and our hearts.

I was somehow a female Archangel—the one and only. I had stopped the dark angel from annihilating us. I would go forward to an unknown future, with Dierdre but without Jonathan. I prayed that I would somehow see him again and find out why I was what I was, and if my death was just destiny, predetermined for this outcome. I believed it was. Destiny.

Jonathan had not met my eyes. He did not move, frozen in place, existing only in the middle world we left behind. Gone were the images of the worlds I had known, ceased were the pain and fear we had all experienced, but forever etched in my heart were the memories of the most mystical place in existence and the only man, through eternity, that I would truly ever love.

About the Author

Elizabeth Eckert has four wonderful children and Saint Bernard dogs. She's a psychology student, a dog trainer-in-training and enjoys spending time with her family and studying people, dogs, and the wonders of the medical and neurological worlds. Nature, writing, and laughter bring meaning to her life.

She's currently writing the sequel to When I Died.

Printed in the United States
By Bookmasters